D1549894

EXTRATERRESTRIAL
ENCOUNTER

EXTRATERRESTRIAL ENCOUNTER

A Personal Perspective

Chris Boyce

With a Foreword by

A. E. ROY
Professor of Astronomy,
University of Glasgow

DAVID & CHARLES
Newton Abbot London

British Libary Cataloguing in Publication Data

Boyce, Chris
   Extraterrestrial encounter.
   1. Life on other planets 2. Interstellar communication
   I. Title
   001.5′0999   QB54

   ISBN 0–7153–7634–9

© Chris Boyce 1979
Foreword © A. E. Roy 1979

All rights reserved. No part of this
publication may be reproduced, stored
in a retrieval system, or transmitted,
in any form or by any means, electronic,
mechanical, photocopying, recording or
otherwise, without the prior permission
of David & Charles (Publishers) Limited

Typeset by Ronset Limited, Darwen, Lancs.
and printed in Great Britain
by Biddles Limited, Guildford, Surrey
for David & Charles (Publishers) Limited
Brunel House Newton Abbot Devon

Contents

CONTENTS

List of Illustrations

(*All by Gavin Roberts*)

A Note on Abbreviations

In the literature concerning extraterrestrials and encounter or communication with them a number of acronyms have been derived which are peculiar to this and closely related fields. A number of these acronyms are used in this book; others have been created specifically for it.

AML   advanced mental lifeform
CETI   communication with extraterrestrial intelligence
ET   extraterrestrial
ETI   extraterrestrial intelligence
ETL   extraterrestrial lifeform
ftl   faster-than-light
LGM   little green men
RBH   relativistic black hole
SETI   search for extraterrestrial intelligence
SF   science fiction
vN   von Neumann probe

For the Mitchell Clan: Ken, Marion, Jennifer, Gregor and, of course, Christopher.

Foreword

This is a timely book. The idea that Man may encounter, directly or indirectly, other intelligent species in the Universe in the foreseeable future is one that people today are taking more and more seriously. Among these people are astronomers, biologists, biochemists, philosophers and linguists. Conferences are being held to discuss the problems of looking for evidence of extraterrestrial intelligence and to devise ways of communication with it should it prove to be still around and interested in acquiring pen-pals. Detection programmes have been initiated at a number of radio observatories to look for life; the Viking spacecraft to Mars carried laboratories to do likewise; other space probes such as Voyager carry messages 'to whom it may concern' providing descriptions of mankind's planet and his choicest intellectual and cultural achievements, inviting any intercepting aliens to come visiting. So this book is timely in the sense that it deals with an idea whose time has come.

It is timely also in another sense.

Many people open to the idea that we may not be alone in the Universe tend to think that any future arrival of aliens or communication with them by radio would be basically an exciting, profitable and stimulating encounter with an intelligent species not too drastically different from us in appearance, biochemistry, thought processes, drives and technology. Certainly it would be potentially highly dangerous in its scope for misunderstandings and conflict; certainly the future of the human race would be altered dramatically; but the resulting intercourse in science, technology, culture, philosophy and commerce would ultimately benefit both races, human and alien. This book is a timely argument that such a viewpoint is almost certainly naïve, perilously so, being based on a too-shallow assessment of the situation. The size of the Universe, its make-up and its age compared to Man's duration upon this planet make it almost inevitable that the encounter, when it comes, will produce consequences far beyond our wildest imaginings.

I am a person who has been intrigued by the possibility of extraterrestrial life for more than thirty years, who has read and enjoyed science fiction for many more, who has contemplated with bemusement the growth of the UFO movement since the last war, who remembers with affection and appreciation *Star Trek, 2001: A Space Odyssey, Close Encounters of the Third Kind, Quatermass* and a host of other movies stretching back to the days of Flash Gordon and ranging in quality from the sublimely good to the unbelievably bad. I am also an astronomer by profession.

Having read Chris Boyce's book I am happy indeed to have the privilege of contributing this preface to it. The book says a lot of things about the extraterrestrial encounter that seem to me to be well worth saying, things that are relevant, exciting, salutary and thought-provoking. This is a book for anyone like me who has at any time looked up at a sky displaying myriad stars scattered like icy fragments over the dark bowl of night and wondered if life was there and what forms it might take.

Archie E. Roy,
Professor,
Department of Astronomy,
University of Glasgow

*Whoever starts out toward the unknown
must consent to venture alone.*

ANDRÉ GIDE, *Journals* (May 12th, 1927)

Prologue

Let me tell you how the Universe was discovered.

It happened in the late nineteen-forties. A small boy and his father were returning home by tramcar from a spectacular exhibition of gee-whiz technological goodies held in the city centre. The child's mind was having spasms of delight at memories of jet aircraft, flying rockets and other promises of the brilliant future which lay beyond the waning pall left over Europe by World War II.

The car shambled noisily to a halt. The tweed-coated father stepped down onto the cobbled roadway, turned and swung his boy in a high balletic arc from the platform. The boy howled with glee. He had become one with the aerial pantheon which currently blasted bolts of ecstatic imagery through his never-distant daydreams.

He was a pilot. He was flying the first jet-powered Spitfire. He was anywhere but crossing a rain-splattered Shettleston Road in the East End of Glasgow.

Abruptly the rain stopped.

A divine revelation of colour suddenly flooded into the world around the father and son like blood pumping back into a numb limb. An arm of sunlight, thick and yellow against the overcast sky, had pushed through the clouds and slapped into the roadway.

The boy stopped and gaped. He was unsure of what was happening. He resented the intrusion into his private drama but the spectacle was totally novel and exciting to those young eyes.

The fallen rain was caught as glittering veins laced among the slick cobbles. The boy shook his head, not quite able to place all this into his scheme of things. Then a show of some magnificence, his first rainbow, drew his attention skywards. The luminescent arch grew ever more solid as he watched; then a ghostlier shadow arch began forming outside it.

Now the boy was almost afraid.

He turned to his father, and saw up beyond the man the breach through which the Sun was bursting. It looked like a hole, a hole in the sky.

'Daddy,' he said very seriously, 'What's above the sky?'

'Er, above the sky?' muttered the man. 'Well, there's space.'

'Space? What's space, daddy?'

'Oh, that's what has the Sun, the Moon, and all the worlds and the stars in it. It's very big. It goes on forever.'

'Forever!'

New and vast perspectives began their slide through the young mind. The whole street scene clothed itself in new meaning, in an image which would dominate that child's imagination thereafter.

The street, dazzling in its countless sparkles of reflected sunlight, became a starfield, infinite, cold and brilliant, stretching away majestic into an inconceivable void. And in there were worlds upon worlds. And upon those worlds were . . . what?

Actually there is no way I can really communicate to you what that felt like. It was explosive: the kind of mortal earthquake that sends shockwaves tremoring up and down the rest of your life. The point is that it was personal. It was a private thing, an individual experience.

There is a reason for opening this book in this particular way. What I write in the following pages originates from my individual viewpoint on the Universe. That viewpoint is not a mathematician's, an astronomer's, or a philosopher's: it is a science-fiction writer's, and one SF writer's in particular. It is mine.

Science-fiction writers have a peculiar qualification for writing about extraterrestrials: they probably spend more time on pure speculation about the nature of ETs as beings and civilizations than anyone else in any other occupation.

'Extraterrestrial encounter.' When the phrase is mentioned there is usually a chorus of groans mixed with chuckles. It seems to be a topic which few people take seriously. They seem to regard the subject as belonging to the fantasies of children and intelligent but quixotic adults or practical scientists who really should know better.

Why is this so?

Complex life-associated molecules seem to be found in deep space on an almost daily basis now. Yet still I see that look crossing the faces of people who listen to me as I answer their query, 'Do you really *believe* they're out there?' It's almost as if I am admitting to having seen fairies turning bananas into sports cars or being a founder member of the Flat Earth Society. As I talk to them I see the muscles of their faces undergo a subtle change, a glint of suppressed humour, a flicker of incredulity. There follows what is the nearest thing to telepathy I have known. The thoughts come across as if they were shouted at me: '*My God, they weren't kidding. He really does believe it!*'

There was a time when the conflict of attitudes brought about by innovations could be happily absorbed by society. This time has certainly passed. As Alvin Toffler argues[1] we are moving through a period when we must now prepare ourselves for the impacts innovation will deal both our society and each of us as individuals within that society.

Programmes concerning the search for and communication with extraterrestrial intelligence have been under way for over twenty years now. Debates concerning the advisability of conducting such programmes have been going on just as long. If this is so, why is the general public so sceptical?

Could it be a matter of exposure? Does the average person equate a contact, an extraterrestrial encounter, with the infamous *War of the Worlds* broadcast put out by Orson Welles? Is any discussion of life on other planets to be considered within the framework of *Star Trek*? Has the broadcasting media beguiled those around us into seeing alien intelligences only in the terms of hoax or fantasy?

If this is so we must begin a crash course of re-education very rapidly.

At the time of writing there is a proposal that the United States' National Aeronautics and Space Administration be funded some twenty-one million dollars for a search programme,[2] and the Soviets plan to place a satellite in orbit to further their studies in this field.[3]

If our technology continues to become ever more refined and we maintain a growing interest in searching for other intelligences the chances of us making an encounter over, say, the next century and a half must become quite impressive, if there is anyone there.

My assumption in writing this book, my personal belief, is that someone is there, or was or will be, and that at some point we will become aware of the fact as a piece of hard irrefutable data. If I can make a private moral bet with myself I will say that this will take place within the next fifty years. When it occurs, the cultural impact may well have a potential dwarfing that of any remotely similar happening in human history. That is why we must prepare for it now.

How do we prepare? We are dealing with a completely unknown factor, the extraterrestrial. How can we possibly come to terms with an eventuality the very nature of which is completely outwith the realms of mankind's knowledge? Obviously there is no way we shall be able to write a manual which will cover everything we should do in every conceivable situation. The permutations of imaginable encounters are probably limitless and, when we do effect a genuine encounter, there is one thing that is certain about it: it will contain elements which we never did and possibly never could imagine.

Before we abandon the unknowable alien entirely we should investigate any method of probing its nature. At present our only tool is speculation. If we bear in mind that any speculative being which we raise in our imaginations will at best only partially resemble the ET, then each of us may speculate. Each of us *should* speculate. Once the sceptics themselves begin to speculate we are in a healthier environment. Then, once everyone realizes that even the richest most probable imaginings are little better than smudged sketches of what the reality is going to be, we begin to adopt a social posture which is ready to accept an encounter.

Let me go into this in some detail.

It is important that as many people as possible begin to consider what the consequences of contact may be for them, individually and as members of society. Certainly the imaginative processes involved will entail masses of wild analogizing. As the nature of the ET is unknown the only solid boundary to the imagination is the very fact that the ET is a complete mystery to at least most of the human race.

Even so, we must approach it, and to do so we must be more fanciful than we would like ourselves to be in our modes of thought. We must create the alien's image for ourselves because we cannot see it. That being the case I want to take a look at mental creativity for a few moments.

I am convinced that we create by finding new and striking analogies between elements in our known world which have never been related to one another before. When we are faced with a problem we extract those elements from it which we can relate to and attempt new relationships with the problem until we understand it; solve it; grasp how it interacts with the rest of our scheme of things.

Let us take the fictitious case of Dr Duck and Dr Goose. Duck is puzzled as to what makes the Sun shine. One chilly night he is ruminating on this while sipping his hot tea and gazing into the fire. Suddenly a eureka bulb lights up in his mind. *Of course! The Sun is made of blazing coal.* He rushes into his laboratory and starts work on his hypothesis. Knowing the size of the Earth, that of the Sun and the distance between them he begins work. Sadly his results show that his hypothesis is utterly mistaken.

Seventy years later Dr Duck's grandson Dr Goose has a similar brilliant revelation while he is working on nuclear physics. He rushes into his lab when the idea occurs that the Sun may be powered by the fusion 'burning' of hydrogen into helium. His theory stands up. Where Duck collected only obscure references to himself in technical journals, Goose collects his Nobel prize.[4] Both men were trying to explain the unknown in terms of the known.

Both men operated by use of analogy.

Both recognized aspects of one thing which they thought could be applied to another. They saw parallels.

We tend to operate this way when faced with a mystery. We pour analogies like liquid wax over the object of our inquiry and then build models from the mould we have made. If the model is unsatisfactory we use new wax, new analogies.

With regard to the ET it is the exercise itself which is important. The continual construction of models and their disassembling and restructuring in novel ways with original elements—this is the essential. The imagined ETs are secondary. Each of them must be profoundly flawed.

The reason for this is the difference in the position of the ET problem and those who wish to tackle it. Arthur Koestler writes:

> To put it a different way: solving a problem means bridging a gap; and for routine problems there usually exist matrices—various types of prefabricated bridges—which will do the job; though it may require a certain amount of sweat to adjust them to the terrain.[5]

The ET problem is complicated by the complete lack of adequate 'bridges' and any recognizable 'terrain'.

What terrain does exist? Let us regard the whole corpus of human experience as the Continent of Certainty. Then we guess that somewhere out there in the infinite Ocean of Mystery exists the Island of Probability: we do not know anything about it except that we believe that it is there, somewhere. Our task is to provide a variety of specifications for the bridge to that island.

We can attempt this only by employing abstractions. These are quite literally not capable of being perceived by the senses. In this sense every ET is an abstraction to us. We should bear this in mind because the abstract has a particular quality of which we must remain conscious whenever we deal with the aliens we conjure from our imaginings: the abstract never includes all that can be said about the real; it represents an element or group of elements which have been selected from the real and held independently to represent an aspect of the real. This gives us an indicator.

Any ET we imagine is by the definition of abstraction incomplete, no matter how detailed and convincing we make it appear; consequently no amount of inference from what surrounds us on the Continent of Certainty can tell us everything about the Island of Probability. It may tell us something. The problem is that we cannot know which speculated particulars of the Island of Probability are accurate or how important they are in relation to one another or to the Island's unknown elements.

This is where we come to realize that our bridge-designing, our model-making, can be useful only as an exercise. A model relates to a system the way that a map relates to a geographical area. The only real systems of which we are aware lie within the Continent and thus we are building there mock Islands and mock bridges. The more we do this the better we become at bridge design. The more varied and difficult the mock Island's terrain the higher grow the chances that we will build the correct type of bridge when the true Island is sighted.

The great problem is dogmatism.

A dogma of ET types evolved from theoretical model-making could be disastrous in the event of a real encounter. Unfortunately, we humans are prone to dogmatism, and it is so easy to confuse the imagined with the real. Jean-Paul Sartre says:

> The two worlds, real and imaginary, are composed of the same objects: only the grouping and interpretation of these objects varies. What defines the imaginary world and also the world of the real is an attitude of consciousness.[6]

That attitude is always to be regarded as suspect where models of the ET are discussed. We must never, for example, assume that the apparent comprehensiveness and/or complexity of a model enhances the prospect of its validity. It should be carefully instilled in all who speculate upon the nature and results of encounter situations that we are performing two limited actions when we make our analogies. First of all we are working with the shakiest form of deductive reasoning, infamous for its capacity to yield invalid conclusions from solid premises. We must also consider the factors which we appoint consciously or unconsciously as those from which we will analogize: we must consider them sceptically because, as I have said, they all originate from the Continent of Certainty.

So where do we SF writers fit in?

We deal out images like cards and each is a story, a dynamic thing. Sartre defines the image as 'an act that is directed towards an absent or non-existent object as if it were an actual body'.[7] This brings me closer to how I see good SF handling the problem.

We want to make the act realistic, exciting and a challenge to the imagination of both the SF reader and the average non-believer in the street. We want to make it seem possible. T. S. Eliot wrote:

> *Between the idea*
> *And the reality*
> *Between the motion*
> *And the act*
> *Falls the Shadow*[8]

At our best we are trying to penetrate that Shadow.

There is already a vast body of speculative fiction dealing with encounters. Literally hundreds, if not thousands, of established, budding and nascent SF writers deal with 'aliens' daily. They have done so, if not in such numbers, for at least a century now, certainly years before H. G. Wells' first Martian missile crashed down upon Horsell Common in 1897.

How much credence should we give the idea that SF writers genuinely attempt to simulate non-human intelligence and civilization? Naturally, being 'one of the gang' I am biassed here to some modest degree. Certainly most SF writers are in the entertainment business and they are in that business to make money which supports wives, children and roofs over the various heads. They are also, for the most part, writing for what they see as a discriminating readership which knows that there are no two-hundred-metre-high methane-breathing swamp giants on Venus and that you cannot travel faster than light in the Einsteinian Universe without a handy 'spacewarp' or 'hyperdrive'.

The trouble with encounter scenarios in SF is that the worst treatments are given the widest circulation.

By and large, the examples which appear on screen are simply trash. In the case of television the scriptwriters concerned are often the same team who work on the opera circuit; i.e., soap, cop and sixgun operas. These writers are highly skilled professional people who can rattle out, alter, scrap and rewrite a smooth custom-built screenplay, precisely to order, and have it on the producer's lap by or before deadline. To the TV companies they are invaluable. For our purposes they are truly worse than useless because they tend to perpetuate the less savoury aspects found in the worst, and some of the best, of the old 'pulp' SF stories.

There is a strong accent on violent conflict and human chauvinism is always rampant. Carl Sagan, the space scientist, uses the concept of chauvinism in the sense of 'the assumption that life elsewhere has to be, in some major sense, like life here.'[9] In the great garbage-can where most mass media SF belongs, human chauvinism also means human superiority. The aliens may be more intelligent than we are but, if so, they lack our emotional capacities and our frequently startling powers of intuition. Naturally their centuries-long technological lead is no match for such capabilities. The result of this is that the alien is defeated and destroyed. Despite the chauvinism there are few instances in SF of books or films ending with a negotiated peace treaty. Human reality seldom intrudes into the dreamworld of human chauvinism.

In Appendix 1 I have compiled a selected checklist of movies which deal primarily with relations between Us and Them. A quick glance down it will show that most film-makers involved regard violence, often straightforward warfare, as being the kind of relationship between us and the ET which the public finds most acceptable. Where outright guns, bombs and death rays play a lesser or even non-existent role there is generally some element of menace. The beautiful extraterrestrial may turn out to be a vampire, for examlpe.[10] Alternatively, the invaders may be psychic parasites which take over or absorb human personalities as was the case in, respectively, *Quatermass II* and *Invasion of the Body Snatchers*.[11]

In the latter film the final scene is open-ended: the invaders just may manage to take over planet Earth. This is exceptional. Generally the intruders are well and truly hammered in the last reel and the human race survives having learned to 'Keep watching the skies'.[10]

Is the ET simply a handy vehicle for quasi-gothic melodrama and pseudo-technological pyrotechnics? Evidently there are those who think so. The problem is, why do they think so? Is it something about the way we react to the very concept of a non-human intelligence at least as advanced and complex as our own?

Do I hear someone shouting 'Fear of the unknown'?

Possibly it is that in part. There is however the lurking suspicion that the ET may do unto us as we have done unto others. In history, where one nation has conquered another there has been extensive and frequently cruel exploitation of the conquered. Colonialism and its present-day economic equivalent, neocolonialism, are fairly well known to most individuals in the West. There is either an abhorrence of the imperialistic overtones associated with them or a variant of the attitude, 'It's good for them because we're civilizing them after all.' In either case the prospect of our encountering superior beings does not bode well.

Wells illustrated this beautifully in *War of the Worlds*. The invading Martians care nothing for what we humans hold dear and are intent only on subduing us. Their technology is incomparably superior to mankind's, their intellects 'vast and cool and unsympathetic'. One of the finest ingredients of the work is its illustration of the fear engendered by this prospect; evidently it shook the author himself to some extent. Mere human invention was useless in dealing with the invaders so, incredibly, the atheistic author had them destroyed by 'putrefactive and disease bacteria ... the humblest things that God, in His wisdom, has put upon the earth'.

Certainly, if the social behaviour of the ET is rather close to our own, this possibility should not be ignored. SF abounds with variations of human social conditions attributed to the alien. As in *War of the Worlds* there are carefully considered non-human intellects with strange social organization. In *The Alien Way* Gordon R. Dickson creates another highly aggressive set of beings, the Ruml, who have a complex society based on kinship and the establishing of powerful bloodlines. In *A Case of Conscience* James Blish's aliens are a threat to Man's image of the Universe, particularly his own place in the scheme of things, as seen from the viewpoint of a Jesuit priest.

There is just no adequate way of referring to the nature of the ET without using SF.

I am not saying that if you want to seriously consider the implications of contact you should write an SF novel. The real extraterrestrials will not be the same as any imagining contained within the hundreds of thousands of stories written about them. There is, from our limited experience, a fair chance that somewhere in that body of literature are a number of pieces which will have a bearing on the problem when it really arises. The problem is so open-ended that it requires an extensive library of speculated possibilities to be tackled at all. This is what SF provides.

Given that, where do we start our study of the encounter? We begin with the other side of the contact 'equation', the terrestrial element, the human animal itself.

Historical Perspectives I

Let us look at Man. He has a wide experience of intercultural contacts. For over five thousand years he has been building cities and creating nation states each with its own distinctive characteristics. The history of cultural encounter begins with history itself, but it should be remembered that there is also an impressive prehistory.

The nomadic and semi-nomadic tribespeople of the Paleolithic period must have met with other groups in their travels. It is even possible that *Homo erectus* of one million years ago had rudimentary tribal cultures which interacted to some degree.[1] To me it appears very difficult for a culture to emerge, thrive and ultimately disintegrate without having some interchange with beliefs and *mores* of other cultures, or civilizations. Sadly, all we can say with much certainty about prehistoric cultural interchange is that it must have taken place. When we move into the realms of our historical past we are on much firmer ground.

There is an argument against the use of historical illustration when considering contact with ETI (extraterrestrial intelligence). It is a valid argument to some extent: we cannot use examples taken from human history since the non-human behaviour of the extraterrestrial nullifies their relevance. This is certainly true when the contact 'equation' is considered as a whole. History is a useful source only when we are examining the human factor.

We know nothing about the psychology of the ETIs. There are reasons, which I will come to in due course, to suspect them of having mutually exotic psychologies. However, to deal with the contact analogies forwarded by both critics and supporters of CETI (communication with extraterrestrial intelligence) it is necessary to assume on their part some quasi-human behaviour. Where possible, however, I will stick as closely as possible to the human side of the equation. In *A Study of History*, Arnold Toynbee[2] points out that there are two extremes of reaction to cultural impact. He called these 'Herodianism' and 'Zealotism' after the Judaic reactions to Rome. In the former, a chameleon-like attitude is adopted in an attempt to become as like the assaulting culture as possible; the latter reaction is to become ultraconservative and eschew everything foreign as a contagion. We should expect both types of reaction to some extent in the ETI encounter. The extent will vary depending upon the nature of the encounter.

We can already recognize the Herodian and Zealot-like psychologies amongst those who make public statements about intelligent life in the Universe. There are those who find the prospects of contact with an alien intelligence highly exciting and promote the idea that we should be actively attempting to discover ETIs. On the other hand, there are those who discourage the idea of drawing attention to ourselves, who believe that we are not yet, and may never be, ready to deal with the problems which CETI might bring.

Perhaps the main fear which lingers in the mind of the individual in Western civilization today is inherent in the process of Westernization itself. Western influence spread from Europe in the fifteenth century to touch all parts of the globe in varying degrees within five hundred years. The expansion has been highly aggressive, if not always violent. In its development the Western way of life has changed in numerous ways. But one element which has remained is the aggressive nature.

What if we encounter an ET culture as aggressive and vigorous as our own? Will our fate be similar to that of the less aggressive societies which have been subjected to the fatal impact of exposure to the West? An analogy could be found in what happened to the American Indians when the European invasions swept across their lands. The Europeans and those of European descent often treated the native population as little more than dangerous local fauna.

The pre-Columbian populations had created their own complex cultures. In Central and South America extensive imperial societies had arisen by the time the Spaniards arrived, notably the Aztec, Maya and Inca. Amerindian civilizations and cultures are now matters mainly for the archaeologist and anthropologist; the influence of European thought and technology destroyed them at the level of authoritative dominion.

The Fatal Impact is the title of a book by Alan Moorehead about the effect of European presences, mainly British, in the South Pacific between 1767 and 1840. The West is shown as destructive, domineering and insular in its attitudes towards the local population. The results of Westernization among the South Pacific's aboriginal people are, if not worse, then scarcely better than the lot of the American Indian. The question which arises here is whether or not these are fair analogies to draw with ETI encounter situations. Few scientists believe that such an encounter will take the form of a spaceship from another planet suddenly materializing above the Earth and landing on the White House lawn, Red Square, or anywhere else for that matter. It is generally thought that we are more likely to receive a very old, perhaps centuries or millennia old, message from a distant civilization with whom it would be impossible to have any other relationship than polite conversation punctuated by pauses decades or perhaps even centuries long.

Currently, faster-than-light (ftl) space travel is regarded as being, if not actually impossible, definitely unfashionable. This slightly less than respectable concept is implicit in the idea of a galactic or at least interstellar civilization as we understand the term 'civilization'. For a society to function as an administrative whole it is necessary that there be some fairly rapid transfer of information between the outlying regions and the administrative centre or capital. By 'fairly rapid' I mean to say that instructions on, say, general policies to be adopted by regional governments might arrive from the capital once every five or six years at the outside. The problem, of course, is that, if the instruction or flow of feedback take an excessive length of time, then the outer regions will become progressively different from the society of the capital because the seat of administration will be unable to react fast enough to regional changes to stop, divert or influence them effectively. By the same token, the 'regions' would differ internally to a lesser extent. A galactic empire governed by the laws of physics as we understand them today would be a temporary affair indeed. Remember, a galaxy is a very large place. A beam of light takes roughly one hundred millennia to cross our own, yet it takes less than one twenty-third of a second to travel thirteen thousand kilometres, which is the polar diameter of Earth. And that thirteen thousand kilometres in human terms would involve the ridiculous image of well over seven million adult males lying in a straight line head to toe.

Most people regard civilizations as fairly cohesive with regard to internal communications; consequently, any interstellar civilization would have to find some method of by-passing the restrictions of physics as they are currently understood.

Let us then assume that we do physically encounter a vast, aggressive interstellar society which has managed to achieve ftl travel. We shall assume that they have been attracted by the television and radio riot which we have been showering into space for decades. If they have been anywhere in our vicinity since mid-century, there is a fair chance that they have heard us. By now we have surrounded ourselves with an incredibly immense sphere of commercial television signals. The TV sphere's radius is currently about thirty light years and growing at the speed of light.

It is difficult to think about distances in terms of light years, so think of it in terms of the volume of our world. Earth's volume is better than eight hundred million cubic kilometres. The TV sphere expanding out from us among the stars is some ten thousand billion billion billion times larger than that at present. By the turn of the century it will be almost three times as large again.

So Earth glows like a radio beacon for the ETIs inside our TV sphere. What happens when their spacecraft appear in the skies above us?

We can be fairly certain that, if they are as aggressive as Western society has been for the past five hundred years, we could find ourselves in trouble, particularly if they regard us as local fauna to be classed along with the other apes. It is generally assumed that this is and always was the normal attitude adopted by the European towards the aboriginal inhabitants of lands he moved into. But is this a distortion of what happened?

Most human beings seem to treat other human beings who are markedly different from themselves racially and culturally with mixed degrees of curiosity and caution, if only at first. This is a perfectly understandable response. Caution is always a factor which emerges when an individual or group is dealing with a new situation.

Regarding the strange person or group with hostility is liable to arise where they compete for limited resources, or threaten the social aspirations of the observing individual or group. The lack of compassion between conqueror and conquered has never been exclusive to intercultural or interracial conflict. Any brief glance at the history of Europe alone is sufficient to demonstrate this. Western society is aggressive. It is also expansionist and economically growth-oriented. The difference between it and aboriginal society is more a matter of kind than of degree. The aboriginal is tradition-oriented to a great degree whereas Western societies are geared more to rapid change. The aboriginal society is thus a poor analogy to use when speculating on a cultural impact requiring change on our part in order to survive culturally. In consequence it is difficult to parallel an ETI contact along North Amerindian or South Pacific lines.[3]

The Inca, Aztec and Maya civilizations, on the other hand, were far from aboriginal. They were socially complex, highly enough organized to have dealt with the Spanish invasions quite decisively before the invaders obtained a secure footing in Central and Southern America. So what happened here?

Unfortunately, it needed more than organization and technology to combat sixteenth-century Spain.[4] There was a cultural difference which worked heavily in favour of the Spaniard. Spain was expanding at a time when European civilization was exceptionally vigorous. Dreams of wealth and conquest obsessed the adventurers sailing westward, who had more often than not sunk a great deal of personal wealth—as well as that of their sponsors—into the expedition. Consequently, it was critical to return across the Atlantic with, at the very least, promise of great fortune.

The Aztec did not realize that he was dealing with people who had the mentality of gamblers and the cravings of the power-hungry. It was the audacity, ruthlessness and technological sophistication of the Spaniard which proved overwhelming. The Inca, Aztec and Maya cultures were destroyed because the invader wished them destroyed. The natives in the upper levels of their social hierarchies adopted a Herodian attitude and moulded their lifestyles into an acceptable quasi-Spanish format. State religions were officially replaced with Christianity and the peasantry were enslaved. Of course, a great deal of the Herodianism was merely the application of what was hoped would be an acceptable veneer. Many of the old beliefs and customs were still in existence but somewhat disguised. In fact, some village communities outwith the invaders' ambience continued much as they had before the Spaniards arrived.

Spain wanted empire and gold. The indigenous political structures of their new lands posed a threat to stability of rule, and consequently had to be demolished and replaced with true colonialism.

If we were to run up against an equally ruthless, cruel and exploitive space-going community which was at least as advanced compared to us as were the Spaniards to the American Indian civilizations, then we too would be in trouble. There are, however, some features which make this appear a less likely possibility than the ETI being comparatively non-aggressive. The first of these is the very nature of contact, of how we currently envisage an encounter.

It is because we are expanding our presence actively out into the Galaxy that ETI encounter is becoming increasingly possible. It would seem from examples of historical encounters that it is the expanding society which initiates these contacts. If this is so, then we can expect the encounter to be brought about by our own efforts.

We may meet another society, but can we expect them to be as expansionist as we presently are, from our historical experience? The possibility of an interstellar invasion fleet, a bug-eyed Cortes in command, arriving in close orbit about the Earth is scarcely more likely today than at any point in our past.[5] The chance of anyone invading a country just at the point when it was undergoing an explosive aggressive situation itself would seem a fairly slim one. If, for example, Pizarro had landed in Peru about a century earlier, when the Inca empire was expanding rapidly, he would have met with a vigorous, conquest-oriented enemy and his victory would probably not have come about so easily, if at all.

If we do make an encounter it means that, evidently, there are other intelligences out there, and this entails the probability that they are already experienced in contact. So much depends upon the rate at which 'civilizations' emerge in the Galaxy and how long they continue to exist. The chances would seem to be against these factors dictating the mutual encounter of two civilizations in exactly the same condition of sociocultural and technological development.

As we are at the very start of our contact programme we are thus likely to encounter an older Intelligence wise to the fact that emergent lifeforms among the stars can be aggressive. Doubtless it will permit us to make contact only when it feels we have matured sufficiently.

What if this is not the case? The aggressive processes of Westernization have been continuing for some five hundred years and show no signs of abatement. Might we not have to face an ETI which has been involved in aggressive expansion for thousands of years? This would be the ludicrous equivalent of Spain's Conquistadores discovering not the Inca and Aztec gold but the much more alluring glitter of twentieth-century North America.

Above all it is this kind of cross-temporal contact which has relevance to our situation. We must realize that coming into contact with a society at once as vigorous as and almost certainly more complex than our own has as much relevance to time as it has to space. As we can talk with any degree of confidence only about human societies, we must think of ETI encounter in the light of our present civilization communicating with representatives of a descendant civilization some centuries hence.

It would be more appropriate, accordingly, for us to investigate contact between our society and others where the impact of the West was felt as near to the present as possible and where the societies concerned were comparable in cultural stability, aggression and social complexity. Possibly the best model would then be that of the Japanese over the past one hundred and twenty-five years. From being a decidedly insular feudal society in the middle of the nineteenth century, Japan has transformed herself dramatically into a potent global force economically and politically.[6]

The attitude of the United States in forcing open the channels of communication between Japan and the rest of the world is very important here. The intentions of the United States were not colonial, at least in the traditional sense, although perhaps neo-colonial in the modern sense. They arrived in 1853 with a moderate show of military might; later they returned to Japan with more ships, carrying a greater complement of soldiery. They were, however, not interested in military conquest.

Had the United States chosen to manufacture an 'incident' it would have been fairly simple, the attitudes of the Japanese being almost pathologically xenophobic. A fair-sized military expedition, particularly one with good logistic back-up, would possibly have crushed the native armies of the islands as they then stood. Considering the renowned spirit and resourcefulness of the Japanese warrior, this statement may seem rather extreme, but there is evidence for it. The evidence is relevant to this investigation of cultural contact.

The traditional warriors of Japan were a hereditary class, the Samurai. These men were extensively trained in the martial arts and lived by a strict behavioural code which governed even the minutiae of their everyday lives. It was only to be expected that when, due to the processes of Westernization, it was decided to disband the Samurai and replace them with an army drawn up along regular European lines, they would object.

The predictable outcome was a showdown between the newly trained force of conscripts and the Samurai. The conscripts won decisively. A Western-style Japanese army defeated the Samurai just as convincingly as the Conquistadores defeated the American Indians. It is the presence of a Western format which both instances have in common.

What factors contributed to Japan's rapid assimilation of Western cultural elements? To understand this we must look at the country existing, not in isolation, but as part of the Far East which from the intrusion of the Dutch traders in the fifteenth century was faced with the 'Western Problem'. The Japanese initially opted for a 'Zealotist' solution by cutting off all but token contact with Europe in 1639. This was done because they did not like what they saw of the exploitive actions of the Europeans. They also had serious doubts concerning the religious, moral and hygienic aspects of the intruder.

This placed them in a situation where they were aware of the West but deliberately ignored it. It was an outside, an 'alien' presence, which stayed outwith the orbit of their own affairs. Although we cannot be certain that a contactable ETI exists, the nearest situation analogous to our belief that it does exist, our blind awareness, is probably the two centuries of Japanese isolation. The main differences are that our isolation is not self-imposed, and that the Japanese awareness was not blind.

Although cut off from the West, news of what was happening in other parts of Asia was obtained quite readily. So, when Perry arrived with a demonstration of nineteenth-century US naval power, the Japanese were quick to realize the consequences of ignoring it. They knew of the British presence in India and the Dutch and French intrusions around the East Indies and Indo-China. They were well aware that they would have to react decisively. Ironically, the Spanish conquest of the Philippines in the late 1560s and the Dutch conquest of Formosa in 1624 had been contributory factors to the isolation policy. It was fairly evident that, if they did not rapidly catch up with the West in military terms at least, then they too were likely to become a subject nation at some future point.

However, it would be wrong to say that Japan was caught totally unprepared. In the earlier part of the nineteenth century there had been some very discreet studies of Western knowledge, particularly in the sciences, by a number of scholars who eagerly, and secretively, learned what they could from the very few Dutch traders who were still permitted to enter the port of Nagasaki. Study of Christianity was banned, but not the pursuits of Western medicine, astronomy,

1 Members of an archaeological dig painstakingly unearth alien artefacts left long ago as a 'cache' for Man to discover (see page 84).

2 A manned NASA deep-space probe encounters a crippled airship of alien origin in orbit around Jupiter (see page 68).

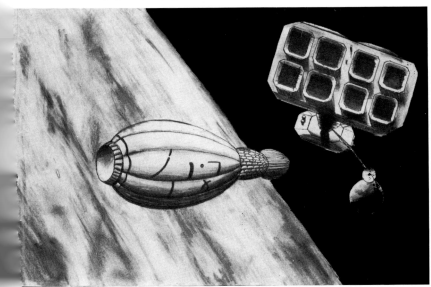

3 A von Neumann probe in the process of construction in orbit around Callisto, the second largest of Jupiter's moons, by men using special handling machines (see page 113).

4 A von Neumann probe in Earth orbit.

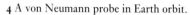

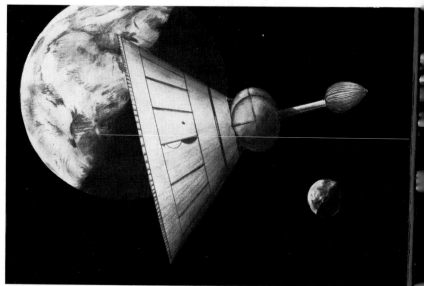

chemistry and military science. These could be followed, but only privately.

There is an analogy here with CETI, which is the scarcity of information we have available, the smallness of the band who are studying it, and its difficulties in being given official credibility by the scientific establishment. The amount of information about chemical evolution and the presence of complex molecules in interstellar space has grown rapidly over the past ten years. It has even been postulated that the thunderstorm-ridden lower atmosphere of Jupiter should be thick with the chemical bases for life. We still know very little, but our information is growing, and with it our perspective on Man's place in the Universe is changing.

Likewise, the Japanese were not wholly isolated from the West during the period of seclusion. They had a number of indicators that, although their own technology had remained static, the West's had not. As the eighteenth century drew to a close, more and more of the unwelcome Western ships tried to gain entry to Japanese ports. 1808 saw a British vessel enter Nagasaki on the excuse of pursuing a Dutch fleet. The captain demanded supplies of the populace, threatening to sink every ship in the bay unless his request was met with promptly. Even Perry's traumatic appearance in 1853 was presaged some five years earlier by another American, Commodore Biddle, in an attempt to establish trade relations without resorting to a show of weaponry.

Between 1853 and the turn of the century was the period when Japan reacted to the West. The reaction was as extreme as that of the 1638 isolation edict. By the end of the first decade of the twentieth century, Japan had an army based on the German model, a navy modelled on the British, a French-type legal code, a centralized bureaucracy, a bi-cameral system of government, a Prime Minister and, that ultimate stamp of civilization, an Emperor who was a Knight of the Order of the Garter.

But ending the seven-hundred-year rule of the military feudatories proved to be a painful and difficult process. Again there are indications of parallels which may well prove relevant in an ETI contact, particularly so in the event of the ETI having very little or no knowledge of human affairs.

By 1858 the Russians, Dutch, French and British had followed the American lead of obtaining treaties with Japan. They regarded these treaties as being masterworks of diplomatic ingenuity. True, they had tricked the Japanese into concessions which were unusual,

for example extra-territorial rights over their own nationals in
Japan, which effectively placed the foreigner above state law.
Furthermore, the Japanese could not decide what tariffs they would
levy without consulting the trading treaty partner affected. The
quasi-colonists felt pleased with themselves because they had
ensured that the Shogun, the country's military overlord, signed
the treaties and not the anti-Western Emperor. They believed that
the Emperor's real political power was virtually non-existent. In
this they were correct. What they did not realize was that by this
time the political power of the Shogun was not much more impressive than that of his Imperial master.

Real power was held by a number of noble families. If *they* so
decided the clever-clever treaties could be declared invalid, as they
did not have the stamp of Imperial approval. Unknowingly, the
Western powers had outsmarted themselves. The Emperor wanted
the 'barbarians' expelled as soon as possible. Internal conflict was
inevitable.

Again the 'Zealot' and 'Herodian' elements are apparent. In an
ETI contact we must expect them to emerge just as naturally as
they always have in human history. May we be as fortunate in the
resulting struggle's outcome as was Japan.

Although the often bloody internal dissent was to continue for a
further eight or nine years, by 1868 Japan had settled its power
struggle in favour of the 'Herodian' group. This consisted of the
younger members of the dominant clans and their Emperor, a
fifteen-year-old boy. The group had all experienced the post-1853
years of crisis as their formative years. They saw the shogunate
established centuries earlier by Tokugawa Iyeyasu as a decayed
structure, collapsing on those within it. Contact, the ultimately
irresistible contact with the West, was the final blow which
shattered it.

The group saw the hierarchical social organization disintegrate as
a means of administration. They also saw something else which
made them decide that the rapid adoption of at least Western
military standards was essential. What they saw was Russia menac-
ing Japan with territorial designs upon the islands.

An ETI encounter demonstrating that the alien civilization's
technology is superior to our own might produce a similar reaction
in us. We may not even understand the language of the beings, or
what their technology is really all about: the very fact that it gives
them capabilities which we do not have would tend to create

anxieties about the security of the human race. Just as the Samurai of the various clans disappeared with the formation of a national army, so might our various national armies disappear with a demand for a Pan-Human Defence force.

There is of course a great problem here. Who administers the power? Who is in control of such a security group? Japan had to create a new social and political structure. We may well be forced to do the same. Japan was fortunate enough to choose among the systems then being used by the more technologically advanced nations. However, in an ETI encounter, where the ETI is naturally different from ourselves in basic psychology and social relationships, it is unlikely that we will be able to do the same. The chances of another creature or creatures evolved along the same lines as the human race meeting up with us must be inconceivably small. Consequently we may have to pull ourselves up to the required level of sociopolitical sophistication by our own proverbial bootstraps.

The chances of our managing this are not as remote as they seem. If the nature of the ETI is quite fundamentally different from our own, and we are in a position to see this for ourselves, the various conflicting powers on Earth would have in common a wish to resist any social influence from the alien culture. In other words, a 'Zealotist' attitude would likely develop strongly with regard to what we saw as weird behaviour in the alien. At the same time we would try to emulate their technological achievements and adapt to our requirements the social organization they used in attaining these. This would entail some degree of 'Herodianism' as well.

Bearing the Japanese analogy in mind, it would seem fitting that, in the aftermath of ETI encounter, we employ as many responsible change-oriented individuals as possible in the process of adjusting to the new pressures. As time goes on it may prove even more fruitful to employ in positions of responsibility those people whose formative years were within the adjustment period, as the Japanese noble families did in 1868.

It would be naïve to imagine that the current world political structure at and immediately after the moment of contact will bow graciously to all this and adapt accordingly. So what is achievable? It seems that unless bloody conflict is to result we should use the political machinery available in each country today.

In countries with a multi-party political system it is not un-common to discover political movements which cut across party boundaries. These often marshal support by forming groups composed of members of the various parties working in concert for their agreed goal. Hopefully, several such groups would form within any given country in response to the ETI encounter with a view to what might be done about adapting to the new knowledge. Should this happen it does seem a natural enough step for such groups to form some links with their cross-political kindred internationally. Thus it appears possible that the political reaction to an ETI encounter would be the emergence of a truly international multi-party system.

This is a necessarily simplistic view of what could occur. To attempt anything more detailed would be foolhardy as it would depend both on the nature of the contact and on the nature of the international political climate. The latter would be further complicated by that of the reactions of individual states at the time.

Not all countries are multi-party in political structure, but even in single-party countries the variety of opinion tends towards the formation of polarized groups. There may be some talk of 'ideological penetration' should any of the groups polarized into existence by an ETI encounter attempt clandestine communication with groups in other countries sharing their opinions. 'Ideological penetration' is also likely to be the kind of accusation levelled against those who advocate that the problem be dealt with on an official international level. This is liable to be a factor where a large and powerful single-party nation might feel able to go its own way in dealing with the ETI problem; such might well be the case in either Soviet Russia or China. The more political parties which a country has at the parliamentary level, the more 'ideologies' which a national constitution can encompass, the less likelihood there is that this attitude against the international movements will be successful.

Would a standing conference of such groups be the body to hold the reins of a Pan-Human Contact Force? Would a regrouping of multi-national/multi-party movements be cohesive enough to tackle this problem? Perhaps not, but this *kind* of shift of political emphasis, if not this particular one, may be needed if we are to cope with the ETI problem in a mature fashion.

2

Historical Perspectives 2

Inappropriate response to the ETI might prove disastrous.

The Opium Wars of 1839—between Britain and China—contributed to Japan's anxiety about the power of the West. Unfortunately, although it gave China a far more evident and bitter display of Western military technology, the Sinic response was comparatively sluggish and directionless.

When change did come the impact on Chinese traditional culture was far more profound than the Amerindian cultures experienced on exposure to the West. Like Japan, China had virtually held the European powers at arm's length for centuries. Certainly she had taken some token steps in the direction of diplomatic relations, notably by receiving a British embassy mission led by Lord Macartney to the Imperial Court in 1793. Sadly the mission returned home having made no headway.

With a surfeit of traditionalist thinking and behaviour, change is impossible except in terms of total destruction. Strangely enough, the same is the case with those who regard only the new as worthwhile (christened 'neophiliacs' by Christopher Booker[1]), to them the past is so much garbage to be junked. A society with the ability to change can accommodate both these extremes; one which cannot suffers the kind of destructive polarization which China underwent.

The old corrupt values of the Manchu government were felt deeply even through the policies of gradual change attempted by the Kuomintang group. Millennia-old traditions generated in an atmosphere of desperation the only kind of change which was possible, the Draconian kind.

China was generally far too concerned with affairs within her vast borders, and the threat of overland invasion, to take the West seriously. Unlike Japan, she could not envisage conquest from the sea as a possibility. Ironically it was the Japanese themselves who came closest to conquering China during the Sino-Japanese Wars of 1894-95 and 1937-45—from the sea, of course.

Almost one hundred years after Macartney's fruitless attempts to establish a British embassy in China, signs of profound change became apparent. Kuomintang, the nationalist party formed in 1891, aimed to modernize the sociopolitical structure of the nation.

A gap of half a century lay between this and the traumatic Opium Wars which ended with the Treaty of Nanking, ceding Hong Kong to Britain and opening the 'Treaty Ports' of Shanghai, Nangpo, Canton, Amoy and Foochow to the United Kingdom, which then proceeded to dominate the China trade.

The difference was that, where 'Herodianism' was triumphant in Japan, China was distinctly 'Zealotist' in its response to the Western problem. There are other historical factors which can be argued as tending to incline the two East Asian states in the directions which they took. From the first contact with Portuguese traders in the early sixteenth century both countries diverged in many aspects of their response. The Chinese regarded the 'barbarians' with a mixture of amusement and irritation. The Japanese were keenly interested, particularly in Western firearms and the lucrative possibility of trade: they went so far as to start their own trade routes, faring as far afield as the straits of Malacca to the south and the Spanish Viceroyalty of Mexico across the Pacific.

Whether or not the Chinese were less naturally disposed than the Japanese to adopt Western ideas and technology, they certainly clung on fervently to their three thousand years of cultural tradition; they were unwilling to introduce any major changes until the structure collapsed, first partially in 1911 and then totally in 1945. By then a new Western format had arrived in catastrophic force. Communism had swept through the nation and was bent on changing every aspect of Chinese life.

Perhaps the two great East Asian nations are today classic examples of the West's two most controversial products, Communism and Capitalism, but Capitalism in Japan is unlike Capitalism anywhere else, as the Communism of China is also unique. Both have created something original from the material supplied by Western politics, technology and *mores*.

From the point of view of culturally surviving a strong impact, the Japanese seem to have been more successful. They seemed to understand the problem, and did something very positive towards reconciling the situation by establishing Shinto as the official state religion in 1882. This provided a strong reference point for the continuance of traditional elements in the lives of the people.

History 'tells' nobody anything. Many of us, however, succeed in reading just about anything we like into it. What we can be sure of is that in the past various civilizations, societies, cultures—call them what you will—have met, inter-reacted and consequently changed. On this basis it is fair to submit that when we make contact with or discover proof of intelligent life elsewhere profound changes will very likely result.

This does not mean that the history of mankind will then be ended; it does mean that history as we know it, of Man existing as the thinking centre of the Universe, will change its nature as it changes its context from the planetary to the interstellar.

The lesson here is that, no matter how tenuous or seemingly irrelevant the ETI contact proves to be, we should respond rapidly. We should adapt intelligently to its presence, altering what technologies or philosophies we can on a careful, selective basis. There should also be provision of a strong reference point for traditional arts and attitudes where we want them to survive.

When the problems of surviving an ETI encounter are discussed there seems to be a tacit consensus that it is cultural survival which is being discussed—and that cultural survival is desirable. What we must bear in mind is that there is also our survival as a race of beings to be considered. We may simply cease to exist as a result of ETI encounter, and not through any aggression on the part of the ETI, but through our own actions. ETI technology and philosophy may be so different that, although they are understandable, we destroy ourselves by employing both or either on a large scale. We may be forced to change our ways profoundly, as has modern China, simply to survive as a race in the Galaxy.

How important then is it to survive as a culture? Surely if species survival and cultural survival are incompatible, we will be intelligent enough to ditch the latter.

An illustration of this kind of problem can be seen in many African nations today. Most African states did not evolve from ethnic pressures but are derived from the colonial exigencies of Britain, France and Portugal over the past few centuries. Consequently, when attempts were made to run nations, whose basic political unit was the tribe and not the state, along the lines of European democracy, the result was frequently chaotic. The Belgian Congo bloodletting of the early 1960s is a frightening and perhaps extreme example of this. Certainly the *coup d'état* and counter-coup have enjoyed some popularity in Africa over the past twenty-five years. With this experience behind them, some political architects currently planning Zimbabwe hope to completely suppress tribalism as a viable force. They will even prevent resident Scots forming a Caledonian Society![2]

African tribal culture will have to go, either forcefully, as is planned for Zimbabwe, or through natural decay. Certainly the destruction of the clan system in Japan was quite systematic, even if the noble families did still retain wealth and some vestiges of political power after the Meiji Restoration. In Scotland, a country deeply involved in the technological dawn of the West, it took place during the eighteenth and nineteenth centuries as a painful process of decaying credibility in terms of political compromise and an inability to cope with industrial society. This similarly affected the aristocratic sociopolitical hierarchies throughout Europe. Of course, the most forceful example of an antiquated structure being destroyed must remain China.

The ETI encounter problem can thus be restated as a human social problem. In our desire to emulate the best of what the ETI has in technology and social/political innovation we may be too hasty.

Let us take as example a case where it is apparent that the ETI will not treat with us because our political machinery is regarded as seriously underdeveloped. If we wish to deal with the ETI we have to change our ways. To change our ways, in a manner which they find acceptable, we have to understand their political systems, and these may be based on a social structure derived from slightly different psychologies.

For example, to enter the Galactic community we may have to live up to our own ideals of democracy, we may have to use the rapid development in computer and ancillary technologies to create a world state in which not only does everyone have a say in everything which is discussed, but everyone is sufficiently well educated to understand everything which is discussed. Furthermore it may be imperative that everyone vote on everything. Compulsory true democracy would be an awesome task and the impact on present day political institutions would be *a priori* fatal.

Like the Asian and African societies, we would be forced to ditch our cherished and, in the case of the West, traditional institutions. These would be replaced by a system which, although seemingly derived from our own political heritage, in fact stemmed from the sociopsychological structure of a race from another star system who had more in common with, for example, Earth's dolphins, whales and other cetaceans than Earth's human beings.

Obviously a political transplant like this is culturally undesirable. What we would have to do is adapt the alien system to suit our own needs. One hopes we would, in time, adapt as successfully as have the Japanese, and that the time of change would be short. It would be pleasant if we could maintain the cultural links with our heritage at least as successfully as has that country. On the other hand, this may prove impossible and we may have to break with our past as severely as has modern China. And, just as that nation has today produced its own singular brand of Communism, so might we all contrive to create a terrestrial brand of 'Galacticism'.

Would this be so tragic?

Either avenue means the end of our kind of world: the end of a world created out of solely human needs. This possibility has not always been apparent. It has, however, been noted by many historians that no one civilization, no single culture, has a divine warranty of permanence. The late Jacob Bronowski expressed it as follows: 'We have not been given any guarantee that Assyria and Egypt and Rome were not given. We are waiting to be somebody's past too, and not necessarily that of our future.'[3]

The truly unchanging element in Man's Universe is, as Sophocles pointed out, the certainty of change. ETI encounter and all its attendant cultural repercussions thus fit well with human experience. If there is tragedy, it lies in the fact that vast numbers of intelligent individuals cleave to the *status quo*, and so must experience trauma when this crumbles. The avoidable unavoided is tragic; recognizing the inevitable need entail only melancholy and nostalgia.

I believe that ETI will be encountered within the next few decades or perhaps centuries; for me that is an inevitability. I also believe that the distances between the stars are so vast that we will have the required length of time to make a well considered series of adjustments at a pace which will minimize potential traumatic impact.

Today we live in civilizations growing increasingly complex both in their infrastructures and in their cross-cultural relationships. Marshall McLuhan's 'global village' is a village in its ease of communication only. Although all nations on Earth would rapidly become aware of the ETI encounter once it was made public, the reactions to it are liable to be somewhat varied, in some cases quite dramatically so.

To return briefly to the 'village' analogy, we should remember that throughout thousands of years of human inter-reaction the village-type social structure remained constant. It is a structure optimized to deal with group relationships and has probably evolved with Man himself from pre-human days when our ancestors moved in packs. It changed little from tribal to feudal within a century of today. Today, however, there are profound changes taking place and many of these emanate from the industrial West.

The group with which an individual identifies himself within his society has always been defined by at least two affiliations: the social and the geographical. From tribal times these would appear to have been fairly clearly defined for most people in most cultures. Even the impacts of acculturation across the centuries tended to leave the bases for group affiliation largely unchanged. Currently Western society is faced with an apparent disintegration of the family as a basic social unit. It is also experiencing 'social mobility', with people changing their 'peer groups' at a far greater rate than before. Kinship and class relationships are losing much of their old homogeneity and rigidity. Commuter suburbs of cities are referred to as dormitory suburbs, as work and leisure take place increasingly either within the teeming cities or within the secluded confines of the home. The village, tribal-based unit has almost vanished. It is being replaced; or, perhaps, as a structure it is changing into a more fluid heterogeneous arrangement of affiliations. As far as the geographical factor is concerned, there is today an increasing likelihood that a high proportion of an individual's friends will be found in another suburb, or even another city, rather than in the next street.

The requirements of aggressive industrialization have forced these changes. In the West we have become acquainted with change and innovation over the past few generations. 'My' world and 'my' values are different from those of 'my father', even more so than the value differences which existed between 'my father' and 'his father'. Such is the nature of continual 'modernization' that we are becoming rather unpleasantly aware of the need to either shield ourselves from it, or grow used to it, or prepare for it through anticipation. As we cannot be at all certain of the nature of an ETI contact, this fluid state of continual acclimatization to growing diversity is highly desirable. So much so that we are probably in a healthier condition today to absorb the effects of an encounter than at any other time in Man's past.

There are other reasons for coming to this conclusion. Science fiction is as I have said riddled with examples of human chauvinism, of the good old Terran triumphing over the merely superficially superior, but deep-down inferior, alien.

Thankfully our 'global village' or communication system is acquainting those who inhabit it with the fact that, far from being inherently 'substandard', other cultures and other peoples have a great deal to contribute to the enrichment of personal and community life throughout the world. Indeed, the past thirty years has seen the emergence of a situation which has not existed for centuries, and which should help minimize the impulse to stigmatize. Once the West was simply the European way of life. The religious and political systems of Europe have appeared to be unchallenged in Western eyes ever since the Polish-Austrian allies thrust the Turks back towards Asia in 1683. But since 1945 we have seen the emergence, not only of Russia as a non-Western superpower, but also of China, and of the Arab bloc operating within and through OPEC. Certainly it can be argued that the political and economic bases of all these are Western in origin; nevertheless, no one in the West would regard any of them as being truly part of the West.

So, currently, there are a number of highly powerful nations or groups of nations each with inhabitants whose daily lives differ quite markedly from those in the other countries. Thus, with the global-political position of any one of these being constantly challenged by the others, none can adopt an attitude of superiority to the rest. In consequence, the Westerner's automatic stigmatizing of non-Westerners as inferior is growing ever more ridiculous in the eyes of the West itself.

If, then, we live in a society optimized towards change, and one in which stigmatizing is becoming less likely, the next step in our ability to handle an ETI encounter is a widespread realization among the peoples of the Earth that the ETI exists. After that has been accomplished, the problem which is unavoidable is the nature of the ETI contact when it occurs. Will it prove palatable?

It would appear to me that any state or society wishing to retain elements of its tradition and culture should not simply be capable of change but should be in the process of change when the encounter is made. The more profound the change process the better.

If we meet up with a culture which, like our own, has no previous ETI experience, there will very likely be enormous problems. The two cultures will inevitably be profoundly different, and both will be deeply rooted in their individual pasts. Assuming both we and they are of roughly equal status in knowledge, the encounter may not be unlike that of the Islamic and Hindu life-styles in North West India during the eighth century. These two religions are not, and have never been, mutually accommodating. It might appear that their followers have always been engaged in mutual conflict, but this is not the case. The Islamic Mughal Empires ruled India brilliantly, with much support from all quarters, thanks to the splendid statecraft of the rulers.

Where there is a great deal both we and the ETI can learn about the Universe, diplomacy will almost certainly come into its own. The human and ETI points of view on the problems which nature poses should encourage us to solve them in concert. This is surely the rational way to regard the 'alien'. The supremely irrational way is to ignore its presence.

Where the ETI has experience of previous contact, things should be somewhat better. They will at least communicate with a prior knowledge of the difficulties inherent in CETI. There is, however, a factor here which grows with the number of ETI contacts which have been made by the other before meeting us. Any Intelligence capable of interstellar communication on a fairly grand scale will probably be wiser than us and more knowledgeable in the techniques of energy handling and application. This would make us keen to study their communications in detail. We would be impressed by their achievement, and would thus view their culture favourably, tending to absorb aspects of it not directly related to the energy handling question. Among inter-relating human societies, acculturation frequently takes place most readily from the materially wealthy groups to the poorer ones. 'Zealotist' reaction tends to be strongest where there is a minimum of material benefit accompanying the cultural influx.

In the case of receiving a communication from a Galactic culture, the chances are that it would be fairly palatable. In order to accommodate the extraordinary varieties of ETI, such a culture would probably have many of the characteristics of the higher religions of mankind. Particularly, it would be flexible in many respects, capable of absorbing many local elements, and generally able to take on native (i.e., terrestrial) coloration. If it is optimized at all, a Galactic culture must be optimized in the direction of diversity, being by its very nature *multi*-ETI.

There will be problems, many of which will not be obvious at first. One of our main tasks will be to identify them, just as a Third

World country moving ahead with a programme of industrialization must understand the dangers of pollution and ecological destruction.

'Galacticizing' Earth will only be complete when we as a race of beings begin contributing to the Galactic awareness as effectively as 'Westernized' Russia, China and Japan contribute to 'Western' awareness today.

We must not be guilty of a failure of nerve. It is absolutely essential that we determine long before the encounter is made that we are going to become a member of the Galactic community, a member with vitality and a new point of view on the Universe which we wish to add to theirs. We must determine to live amongst the starfolk as their peers.

If this is our personal, emotional and intellectual commitment to the future, then human civilization will move forward to the stars simply because that is what it has set itself to do.

3

The Dark Pedigree

Being human is the most natural thing in the world—to us, that is, and in this world. It seems to us that the human mind has achieved some kind of universal thinking standard, a standard which is the truly natural form of intelligence not merely here but throughout the known Universe.

It is an easy trap to fall into. As far as we can peer out into the vast reaches of space the processes of physics seem to be constant. There is a body of opinion that appears to believe that minds capable of perceiving those processes will have a common basis for building up a language for discussion, even if those minds have evolved on the planets of different stars.

The problem here lies in that the structures of the minds perceiving the same phenomena are not directly related to the phenomena under observation. An exploding sun, a nova, is not directly linked with the mind which observes it. Consider a volcanic eruption being viewed by both a buzzard and a giraffe. It would have no connection as an event with the design of the brains which registered it. Both brains are complex, having evolved over millions of years in highly specific ways. The interpretation of the event, its *significance*, would be dictated by the specific brain concerned.

To show how individuated significance relates to experience we need look no further than *Homo sapiens*, ourselves. Indeed it is imperative that we do look at ourselves in this light because thus we build up a little more of the human side of the Encounter Equation.[1] Then to balance out, on the principle that we are average, we attribute a similar level of complexity to the ETI's side.

We are at the furthest point reached in our evolution as a species. The same can be said for all the other species on our planet. The situation is not static because we all continue to evolve as species. We try out new variants, mutations. So let us attribute to the ETI a process which, if not natural selection, is at least as complex as our own.

So far there have been many organic molecules discovered in the clouds which lie between the stars.[2] There is even some evidence from meteors that the very 'building blocks' of life have evolved outside the Earth and should probably appear through known processes on planets similar in a number of respects to Earth in its primordial state.

There is always the possibility that through undiscovered processes mental life can spring fully formed from the chaos. This would open an area of correspondence between ourselves and any other entities with a long developmental history. Sharing this factor may prove a common ground whereupon we might start forming the basis of a closer liaison and perhaps ultimately some profound communication.

Of course, they will not be related to us genetically. My opinion of those writers who would have us believe that we are the result of genetic engineering by an ETI is that they explain nothing and open no new possibilities. Some are sincere, others are not and there is the odd one here and there who appears to be crackers.

If there are differing forms of developmental history then each may well throw some kind of light upon the other, enlightening us, and hopefully the ETI, about the nature of our origins in the Universe.

Evolution, for example, has a number of restrictions as we know of it on Earth. It can and does pass on the mechanisms which permit us to learn from our environment but it does not pass onto our offspring that which we have learned. This process, the inheritance of acquired characteristics, or Lamarckism, may well be present elsewhere in the Universe. Beings designed along these lines would have a great number of physical and mental differences from mankind. Their psychology is virtually unguessable and their rate of acquiring data about their surroundings would be astounding to us. Perhaps even describing them as a plurality is wrong. They may be 'It', a multifaceted organism distributed throughout a vast number of individual bodies.

How would such a mentality view us? Perhaps we would seem like some very odd freak of nature, if It could imagine us at all. To It we would lack the basic qualities requisite for an AML (advanced mental lifeform) and, if It speculated on the existence of creatures like ourselves, we would be dismissed as a statistical impossibility.

There simply are no hard facts about ETI. Reflections on its nature are at best hypothetical. Presently ETI itself is an hypothesis and an *ad hoc* one at that. The very discovery of its existence, should it happen and hopefully it will, is going to change our picture of the Universe. It may well change it fundamentally if alternative forms of developmental history are established.

Should this happen it will hardly be tragic. One of the points about knowledge is that it is relative. Theories happen to explain how various events seem to be inter-related at a certain point in time. They are rough sticks which we use for measuring the world about us and which we throw away when better, more accurate, ones become available.

It is important to make this clear before going any further because, although I am going to talk about evolution for the next few pages, evolution is not a fact: it is a theory. If a thesis is published tomorrow which explains everything which evolutionary theory does and more, a great deal of the following wordage may have to be ditched, or at best revised.

The first point which has been made by numerous writers is that the number of chance events which went into the evolution of Man is so vast that there is virtually no chance that Man as he exists on Earth exists anywhere else in the Universe. A lifeform evolving on another planet would certainly go through an equally vast number of chance events in its development. It would be similarly unique in the Universe. What we have in common with the ETI is our singularity.

What constitutes a human being is more than a mere physical collection of arms, hair, bone, nervous tissue and so on thrown together at random. Like every other living creature on Earth a person has a profound intimate relationship with his environment, which naturally includes all other life coexisting with him. This is not a closed biologically exclusive arrangement. Volcanic action, earthquakes and ice ages, along with a catalogue of other non-biological activities, tend to act on lifeforms rather than be acted upon by them.

When a species can survive most types of natural catastrophe as a species it is well on the way to establishing its own private ecological niche. This is why the delectable little oyster looks and very likely tastes pretty much the same as it did one hundred and fifty million years ago. Every new generation which appears has a few wild variants thrown in by nature. These might be regarded as natural experiments. Most of them fail. In the case of the oyster they have been failing for some considerable time now because the oyster is

5 A von Neumann-type probe departing the Solar System having left behind it an 'egg' (see page 114).

6 An asteroid in the process of conversion into a probe factory by automata from a visiting alien probe of von Neumann type (see page 124).

7 A von Neumann probe gathering construction material from the rings of a Saturn-type planet (see page 114).

about the optimum design for the conditions it has experienced since it snuggled into its particular niche.

Fortunately for us, the various forms of animal life which were the precursors of Man had a higher success rate with their 'natural experiments'. Conditions would change and an experimental model, a mutation, would prove to be particularly good at handling the new set of conditions. This would continue as the various changes took place in the world around them.

One of the crucial factors here is that an evolutionary adaptation on the part of one species would be likely to lead to adaptation in others. If, for example, a small mammal changes from foraging for food by day to night-time, the predator who relies primarily upon it for food will have to adapt. It can either become a night hunter or it can change its design so that it is better equipped for hunting another prey by day.

All this is done through natural experiment. It is a highly wasteful but nonetheless successful process, which has inter-related the development of life on Earth for the past several billion years. This developing inter-relationship within the consequently altering environment is a crucial element when we come to the ETI problem.

If we assume that a system similar to evolution has been responsible for the production of an extraterrestrial AML, we assume that the developmental history of the organism is likely to be about as closely tied to that of its environment as we are to ours. It is not just the ETI which is singular but the environment which has produced it, and the whole sequence of events which has brought that environment into being.

All living creatures that we know of relate to their environment through their activities, their behaviour. This is just as subject to natural experiment as the physical design of the creature. When there is a favourable combination of design and behaviour there is a major change taking place. A bird's wing accurately reflects properties both of the air in which it operates and in the way in which the bird operates the wing. This is an illustration of environment, design and behaviour successfully integrated.

What then about the design of Man? Like any other animal his physical and behavioural aspects are the expression of his relationship with his particular environment. Behind this there is an impressive history of development stretching back as long ago as life itself on this planet.[3] It would seem from this alone that he is closely bound behaviourally with the limitations of his unique environment.

The instructions which each member of a species carries in every cell of its body for the duplication of that individual member are held in the genes; Man is endowed with more genetic information than almost any other known creature. This information is not simply a design blueprint: it is also an operating manual. The basis for the individual's intricate interaction with the environment is coded there.

It seems ironic that the animals which have a greater ability to learn need to have a greater store of genetic information. On reflection this does take on at least a patina of logic. As time passes, animals appear to have an increasingly sophisticated relationship with the world in which they operate. This is reflected in their design. Even the growing sophistication of behaviour is reflected in physical design.

The design of the brain is one of the keys to the recent evolution of Man. This brain, like the rest of the human being, has a pedigree stretching into antiquity. At no point back down that branching path does it even touch the development of an ETI brain. How could it? It was never involved in the equally complex and distinct dynamic changes within which the brain-equivalent of the ETI came about.

Man's ability to modify his behaviour and to learn have come about within the context of the changes on planet Earth. The ETI's abilities will have come about within the context of environment x. Earth and x are quite distinct and different. There will be a great deal which they have in common. There will be common chemical elements and compounds; the physics of x will be the physics of the Universe. However, the abilities which the ETI possesses to appreciate these factors will be differently derived from those of Man. They will be closely linked to the ETI's developmental history within x.

This does not mean that there can be no possible communication between us; but I hope that it does show the problems of CETI to be more complex than many would believe them to be. There is a tendency to attribute through analogy the developmental history of Man, our own pedigree, to the ETI. Analogy can be dangerous. Here it is disastrous.

In one very learned text on CETI there is a discussion about the rise of technical civilizations.[4] It is based on a close study of the rise of human civilizations and the problems which could be involved with a society reaching the technological level. Personally I think that this is entirely the wrong attitude to adopt. Even noting that human civilizations had to undergo a variety of transformations before attaining their current states is hardly enlightening. It tells us nothing about the ETI which we cannot already surmise from its having been subject to a system roughly similar to evolution.

The really dangerous assumption to make is that the ETI is *civilized*. Attributing a corresponding level of complexity to their side of the equation does not presuppose the same or even a similar system.

What is required by any mental lifeform in order to expand the results of its mental activities across the cosmos? Is civilization the only likely route available? If this is so, then the reason why we have yet discovered no convincing evidence of ETI is that we are certainly alone. The sequence of events which led to Man was singular enough without adding to this the further socioanthropological phenomenon of civilization.

This assumption, that ETI will have the behavioural trappings of Man in a number of respects, seems quite bewildering. The very people who discuss it acknowledge on the one hand that Man is unique by virtue of his evolution and on the other make quite a point of illustrating how similar to him they expect ETI to be.

Let us anticipate that the non-human mind expands to the stars, if only by virtue of its electromagnetic excreta as we are doing. Even so, we still cannot assume that the behavioural systems which brought the interstellar urge into being are similar. A skinned rabbit is a skinned rabbit but there is more than one way to skin it.

The ETI will be an AML which has achieved its advanced status by taking a different path from our own. Consequently we will find its mental activity to be rich but bewildering, and certainly exotic. Thus an encounter will not be an interaction between intelligences as we understand intelligence. Our mental structure, which defines our mind's activity and thus what we describe to ourselves as intelligence, will not be the same as the equivalent structure/systems in whatever we meet. There will be two interpretations of the encounter. One will be ours, that which we describe to ourselves using all the information which our mental equipment is dealing with as 'objectively as possible' from the situation. The second will be the alien form, that of the ETI, equally sophisticated but in many aspects differing profoundly from our own. So in some ways the two of us will be roughly akin here to the buzzard and the giraffe watching the volcanic eruption.

A single event will mean different things to us. Our hierarchies of significance are going to differ in many ways. These hierarchies are involved in the very moment-to-moment experience of existence. Consciousness as we understand it may be a mystery to the ETI simply because it is not relevant to it.

The confrontation with ETI or the products of ETI will prove to be a most startling revelation to Man. We should be prepared for a shock. We should be prepared to deal with the possibility that an advanced seemingly technological ETI will be far more different from us in its modes of perception and cerebration than we have so far led ourselves to suppose. For humans the existence of an ETI with what we interpret as a technology indicates that the ETI is a lifeform which understands scientific principles, logic and mathematics; it would therefore seem that here we have something solid for initiating a basis for communication. This belief is fundamentally misconceived in my opinion.

It has been pointed out that there are certain aspects of the Universe which remain constant and are not subject to individual interpretation. This is true. It has been posited that we may use these as a basis for communication. This may well be true in part. The wavelength of hydrogen emission in the electromagnetic spectrum (21cm) has been suggested as a likely one to be used by ETI attempting to communicate by radio across interstellar space. This appears to be reasonable; so far so good.

The problem arises in what follows. We know that thinking is a human activity and science is one of its products. Science is a way of thinking. It is a way of thinking particularly concerned with pushing forward the boundaries of human knowledge. If we pick up a complex signal—say on the 21cm wavelength—and after considerable testing build up a credible argument for it being artificial we are liable to believe that, as a technological source is responsible for it, we will be able to understand it. This is because we think that the source could not send the signal without an appreciation of mathematics and logic, as mentioned above.

Our minds, even at their most abstract, function with reference to the terrestrial connection. We have evolved in such a way that we form symbols and extract significance from the environment. These we combine in what we call relationships. We are good at playing with relationships. Other animals on Earth, particularly mammals, appear to have something of this ability, too. Certainly the other primates have some symbolic conceptualizing ability, as has been demonstrated with chimpanzees who have been taught sign language and the ability to communicate with human beings *via* a computer.[5]

The nature of the problem is better illustrated as we move further away from Man, rather than closer to him, in the world of animal life. The cetaceans, the dolphin-whale family, are well known examples of animals with highly complex and large brains. It would appear, however, that the newer areas of their brain, the neo-cortical forebrain, developed in the sea, a quite different environment from that in which Man's appeared. Zoologically the chimps are our very close cousins whereas the cetaceans are much more distantly related. The zoological and evolutionary distances are illustrated in both the appearances and behaviour of these animals. Accordingly, it should not come as too much of a surprise that we find it comparatively easy to communicate with the primate and extremely difficult to communicate with the cetacean. To reiterate what I said before: the developing brain is a reflection of the creature's adaptation to its changing environment. With a history of millions of years in the oceans of the world it is only to be expected that the mind of the cetacean is very different from our own. This is going to be particularly true of the area of non-genetic information, information which the environmentally produced brains extract as being significant from the Universe as they see it.

In other words *significance* is the key. What is of significance to the whale and dolphin may be of little if any significance to a human being. Those cerebral structures which attribute significance may well have developed similarly in a number of respects, particularly as they are both mammalian brains in this case. This does not mean that the particular significances extracted are the same.

As far as structure and function are concerned we can expect a good deal of correspondence. Similar structures should have fairly similar functions. To use an analogy, let us consider a machine for transporting human beings through three dimensions from one point to another. The vehicle will probably have a shape dictated by the properties of the medium through which it moves. Other properties of the medium will be utilized by the propulsion system and the humans transported within will be protected from the vagaries of the medium by a closed and pressurized environment. This one description of structure and function applies equally to aircraft and submarines. The reasons why these machines cannot operate in each other's environment are fairly obvious. There is nothing of particular significance in the other's environment which allows the one to operate within it successfully.

It can be dangerous to extend analogies but, if we humans consider ourselves to be cerebral Space Shuttles, the chimp is an old turbo-prop aircraft and the dolphins are way down below the waves, nuclear subs perhaps.

This is a better explanation of why the dolphin-whale minds have no civilization than the fact that they are not equipped to handle tools and thus actually create tools. Civilization is a by-product of the *human* mind and an illustration of *human* behaviour. To expect the same from the cetacean mind is naïve. I will go so far as to say that civilization is insignificant to the minds of dolphins. If we wish to come to terms with them there is no point in studying them in 'controlled' environments. These probably have significance only to us, not to them. If we seriously seek to understand we must join them in their world. Long, close observation will not be enough unless it is done within the environment where these creatures interact optimally with the Universe. If we would under-stand them, we must not expect them to set up a dolphin embassy in Sydney harbour.[6] We must establish the embassy, not by the shoreline, which is our symbolic, and thus by definition human-interpreted, division of the two worlds of man and cetacean, but place it where it cannot be ignored—travelling with these creatures, among them through the oceans of the world.

It is important that this be done. Discovering the nature of the complex dolphin and whale minds is a step towards the under-standing of how best we can deal with the much greater problems of understanding mental life in general, our own in particular and that of the ETI, should we ever contact it.

Broadening our concepts of what mentation is and what it does will lead us into many new philosophical realms which are currently great fun for the science-fiction writer (particularly Ian Watson[7]) but little considered by the scientists working on CETI. The simple chain of consequences which I would like them to take note of is as follows: non-human beings have non-human minds displaying non-human behaviour.

The behaviour which is derived from our own dark pedigree has been the subject of considerable debate. It is important that we determine as closely as we can how the human being is motivated, what relevance genetic instruction has in coding behaviour into the brain, and to what extent the structure of the brain itself delineates the boundaries of human behaviour. Civilization, human society and culture are all products of that unique being, Man. We cannot expect them to exist elsewhere in the Universe any more than we can expect Man to exist elsewhere.

What we are liable to discover out there are complex and exotic entities which will appear, at least primarily, utterly bewildering.

4

Come Up and See Me Sometime

If the ETI has a considerable amount of experience in encounter we can expect to be led along the path, if not actually up the garden path, by a being much more subtle than we might initially suppose.

A veteran AML preparing to open relationships with emerging mental lifeforms might regard a knowledge of the contactee's psychological bases to be desirable. Playing around with junior's behavioural building blocks might be one of the general approaches to the whole problem of minimizing and re-channelling violence if it is common amongst the 'youngsters'.

Although we humans readily identify the biological bases of behaviour among other members of the animal kingdom we find difficulty in ascribing this perspective objectively to ourselves. If hubris is the stumbling block then the ETI viewpoint on ourselves should dispel that effectively once their presence becomes known to us.

Remember, to an ETI *we* are the aliens, the bizarre creatures inhabiting a strange and mysterious planet. To them we are the dominant lifeform of Earth, the animals who have adapted most successfully to the stringencies of Nature as present on this peculiar world. We have done this due to the singular nature of our mental activities. Physically an ETI would class us as a primate along with the other apes. The big problem would be in the comprehension of our behaviour.

Obviously if you or I or anyone wishes to become accepted as a member of a community we should adopt the standards of behaviour which are expected in that community. Likewise, if we wish to establish diplomatic relations with another country, it helps to know how one is expected to behave socially; and, should one be expecting a fairly long stay in the country, a more than passing acquaintance with the language is highly desirable. An ETI wishing to establish relations with the human race might therefore wish to understand our biological-behavioural language or 'biogrammar',[1] taking as a reference our fellow animals here on Earth.

Here they would have a tremendous advantage. We have no knowledge of the biological bases for ET behaviour. As we cannot assume that they are similar to our own, CETI specialists should take great care about what they mean when they refer to 'advanced technical civilizations'. In their estimations of the numbers of such societies they frequently refer to the proportion of them which communicate by means of radio or other electromagnetic media. A further factor is the lifetimes of such societies and whether or not many or few may destroy themselves.

The surviving of potential technological self-destruction strikes me as a peculiarly human consideration. If all AMLs are quite distinct in their natures then the nature of the problems which they create for themselves are liable to be solved only within their own sphere of understanding. The concept of a Galactic Agony Column where all cries for help are succinctly dealt with strikes me as being anthropomorphic.

There are those who seek to justify CETI in exactly these terms. How might another AML capable of studying us in something like our own terms regard the human desire to search outwith oneself for answers to internal problems? In this instance, CETI seems like one of the oldest forms of communication with a very advanced lifeform; it sounds like prayer.

We cannot assume that there are 'advanced technical civilizations' or even 'scientists' amongst the other AMLs in the Galaxy. We should not even assume that there is consciousness as we understand it and are aware of it. This would appear to be governed by the limitations of our brains imposed by that genetically governed structure evolved out of innumerable chance events happening over millions of years.

When we are awake, conscious, our brain is in its behaviourally active mode. To use an analogy with computers, when we talk of consciousness we talk of the condition in which the behaviour programmes of the brain are 'on line', available; the functions of the brain structure are ready to be used. This expression of structure is completely absent from all known forms of vegetable life because there is no structure. This does not mean to say that wherever there is a structure of this nature it will be similar to that evolved among animal life on Earth. The structures will differ and the expressions of them will differ.

We can expect no scientists elsewhere and no civilizations in terms which we currently appreciate.

What another AML must have is equally complex structures generating equally elaborate functions, as sophisticated as our 'consciousness' but different. As we have no worthwhile analogies there is no way of realistically anticipating what these may be. What we can appreciate is how *we* might bridge the gap and assume that in the case of some ETIs there will be a function analogous to this.

Should it be able to, it may study the behaviour of Man comparatively with that of other animals which surround us. If it is very different from us it may have to start with the basic differences between plant and animal at the level of single-celled creatures and work laboriously up: this may be necessary in any event, considering the complex interaction between all living beings on the planet.[2] Eventually it will appreciate us within the mammalian and then simian frameworks. On that basis most of the guidelines for our behaviour will have been laid.

They will then understand just how different they are from us and their world is from ours.

Bearing this in mind, how should we react if an ETI does establish contact? We should realize immediately that any response we might make could be either misunderstood, incomprehensible, or anticipated. By 'anticipated' I mean that the ETI would have done its homework and learned our 'biogrammar' before opening the dialogue.

To do the homework thoroughly much of it would have to be done *in situ*, which would mean that the ETI would have to visit the Earth unobserved. Again we return to the unfashionable hypothesis of interstellar travel. We will continue to return to it. Why? Frankly I find the concept of a Pan Galactic Radio Network smacks of CBS or the BBC expanded to Galactic proportions, and equally farcical. Without an appreciation of the 'biogrammar' of the beings to whom a broadcast is being directed, assuming they even find radio desirable in the first place, the chances are that the message to the stars will be at best garbage and at worst frightening. Only through rigorous study of the lifeform concerned, probably at close range, can we (or an ETI) make effective attempts at communication.

At our present rate of technological progress we will have the capability to send a probe to another star system some time in the twenty-first century. Consequently, to talk in one breath of civilizations *thousands* of years ahead of us and the impossibility of interstellar flight is short-sighted.*

Many AMLs throughout the Galaxy may not be interested in— or even able to think in terms of—interstellar travel, but mobility between the stars has probably been accomplished many times in the history of the Galaxy. My own contention is that the motives for the mobility are probably very different from one another and from our own.

I will assume for the purposes of the present argument that an ETI has been present in the Solar System for some considerable time and is studying mankind with a view to making a contact. What form would the encounter then take?

There are aspects of the human being's behaviour which would have to be taken into account, such as xenophobia, aggression, fear, suspicion, hypocrisy, cunning and lying—to give but a few examples. Naturally an encounter would have to be staged in such a manner as to give as many psychological reassurances to mankind as was feasible so as not to trigger, perhaps quite literally, an unfavourable response.

There should be evident, right from the start, some aspects of the ETI which make them interesting to us. If these are not immediately apparent, then other artificial aspects would best be manufactured to engage our interest. The very fact that they are ETIs and have interstellar capability will of course be interesting as such. What I am referring to is the presence of the ETI. If, for example, it is drastically different from us physically (e.g., gaseous or microscopic) an artificial physical form might have to be used until we were ready to accept the truth about the physical reality.

Note, these artificial physical forms would be employed to help the human through the psychological doors of encounter. If this is so, then there is no reason to suppose that the ETI would stop short at partial deception. A wholly artificial encounter designed specifically around the needs of the human psyche would be the ideal for an encounter with us. If they have learned our 'biogrammar' sufficiently well they will know this. If they are the equivalent of a few thousand years, or even a few centuries, ahead of us they should be able to put their knowledge to good use.

For the above reasons I feel that our first encounter could be idyllic from our point of view, *should an ETI know of our presence*

*See Appendix 2.

and be moved to contact us. Not only would our psychology be utilized to defuse potential conflict, it would be applied to further the ends of the ETI. In an encounter we naturally assume that the ETI wants as satisfactory a contact as possible. What that would entail from the visitors' own point of view we cannot say, but, if they wish it to be satisfactory from our point of view, we can hazard a few guesses.

It would be nice, for example, to run into a particularly likeable group of ETIs.

The TV programme *Star Trek*[3] has gone through a fair gamut of ETIs and, however unsatisfactory most of them may be, some indication can be had from public response to the various 'aliens' which kind would possibly be the most palatable to the general public.

There is of course the remarkable dispassionate but caring Mr Spock. What makes Spock appealing? He has an ethereal quality about his personality, a detatchedness which seems to border on the divine. This marks him in many eyes as a superior being. Such a being would certainly be attractive to those who look to the heavens for a measure of divine guidance. (It has been suggested by Jung that the UFO cult is in fact an attempt to replace God with ETI which is superior and benign.)

If a throng of Spock-like beings ('Vulcans' in *Star Trek* parlance) did appear in space ships orbiting about the Earth the reactions would probably be rather mixed. There are those among us who do not like the idea of being inferior to anyone or anything except, perhaps, God. Being faced with a race of superior creatures implies that we are faced also with a superior authority. Would governments feel that their subjects saw the ETI as an ultimate supra-governmental court of last appeal beyond the Earth? They would certainly feel that their own authority was undermined and thus tend to regard the ETI with some measure of hostility. Furthermore, although paragons may be admired it is rather difficult for the rest of us flawed mortals to identify with them.

So if there is, somewhere in the annals of that immensely successful piece of popular television, an indication of how the ETIs may approach us, it lies only partly, if at all, with the Spock image. In fact, one particular episode evoked an immense response to its 'alien' lifeform which had nothing to do with the Spock-type appeal. These creatures were not particularly intelligent at all.

They were called 'Tribbles'. They were utterly lovable. The urge not to harm them, never mind destroy them, was the point of the episode in question, as they were posing a threat to the starship due to their excessive breeding habits.

I think that here we are on much more likely territory for an ETI encounter designed by Them around human psychology. If the encounter cannot avoid being a provocative occurrence then let it provoke this particular kind of response, not just pro-ETI but one which brings out a fondness from us, creates an emotional bridge with the visitor.

The forming of emotional bridges or 'bonds' is an integral part of our biogrammar. It is so basic that it may prove accessible to them. Observation would show it to operate through a vast section of the Earth's animal life, as I have said. They might be able to deduce the detail of human bonding from the general activity.

So what kinds of bonds are there and how could the ETI go about employing them in an effective manner?[4]

The most difficult bond to create for them would be the parental one. It would be a tough assignment indeed for an ETI to somehow persuade the human race that we brought it into being and try thus to elicit a parental response from every one of us. A 'pseudo-paternal' response, on the other hand, might work. This would be to evoke the same or similar kinds of feelings which are brought out in the parental one. This is similar to what the Tribbles did. They evoked a feeling of wanting to care for the helpless 'little things'. The parental response is one which encourages us to protect and cherish the 'child' object which can be either a child or a pet or another human being (who is at least nominally adult).

There is a problem, of course. We are liable to find it difficult to believe that an ETI capable of interstellar flight is a helpless 'little thing'. The technologies involved in such an engineering project are daunting enough for us to regard the ETI as a quite formidable 'little thing', no matter how cute and 'Tribblish' it may appear. Nor will a combination with the Spock image help any. Where the Spock image is aloof, the Tribble image is cosy, and the combination of a cuddly Spock or an aloof Tribble nullifies the effect of both.

The bond will have to be such that we find the ETI acceptable and attractive. Only the most perverse humans want to destroy that which they love. They tend to pursue and make room in their lives for love objects, be they child, spouse, the music of Beethoven or the game of golf.

And the game of golf has to be taken seriously. It is an example of a vast number of human activities which have arisen out of an urge on the part of human males to bond together and form groups from hunting parties to football clubs and all sorts of societies with dark and secret initiation rites. This is the male group bond.

It seems unlikely that this would be used by the ETI in other than a reinforcing context (which I shall come to shortly). It is a trifle surrealistic to suppose that ETIs will suddenly appear in the sky above and make themselves known to us as orbiting representatives from the Tau Ceti chapter of the Masonic Lodge.

Possibly the best strategy for bonding will be one based on sexual lines. This is not as ludicrous as it may at first sound. Enough investigations of human sexual proclivities have been made to show that what a human finds sexually attractive is dictated by that person's emotional development within his or her personal and/or cultural environment. Our sexual appetites, although generated by our reproductive process, are not dictated by it. Happily for us as a species this discrepancy has not proved so far to be of major proportions.

A humanoid designed carefully to accommodate many of the features prized in the sexual aesthetics particularly of the West would prove a most attractive creature. It should also have built in some of the features which evoke the parental urge; these have a good deal to do with the distribution and relationship of the facial features one to the other. Large, dark 'wet' eyes are, for example, generally regarded as being attractive. We seem to find them so in animals, children and other adult human beings—particularly of the opposite sex.

It would be important to make the 'alien' at once attractively human-like but not wholly human in appearance. Were we to encounter another group who were apparently human, our suspicions would be aroused immediately, as we know how unlikely it is that there could ever be another race of human beings elsewhere in the Universe. Consequently, there would have to be differences—and these themselves would make it less likely that we would endow the ETI with the duplicity which we associate with our fellow humans.

Yet these differences, too, would have to be oriented towards making the custom-built being attractive. Give them then the short reddish pelt of a deer, say. Give their facial features doe-like or feline aspects. Make them on average perhaps slightly smaller than a human being, so that they do not overawe *Homo sapiens*. Let them move with a svelte grace of step and speak softly, laugh musically and frequently. Above all let them be able to smile prettily.

This is a rough sketch and many males will disagree about how attractive a bonding object is, particularly in sexual terms. But the general principle is what I want to put across here: giving the bonding element as widespread a shotgun blast as possible so that something is hit somewhere in everyone—or most people—who encounters the contact 'fronts'.

The initial encounter itself is of great importance. The ETI presence should be made known in such a way as to promote a minimum of panic or hasty reaction. I am specifically dealing with the presence of an interstellar spacecraft of non-human origin being introduced into human consciousness. How would it be done?

Coming on it 'cold', as it were, would possibly be a traumatic experience. We would have to be weaned in through acceptable channels. Currently the only acceptable channel is that of the radio signal from a distant star system. Should the ETI decide to incorporate this as a feed-in to the situation in the near future—say within the next fifty years—we would find it quite palatable.

Let me construct a scenario.

Radio telescopes are alerted to the fact that an apparently artificial source has been discovered in a part of the sky thickly populated with likely stars. It is roaring so loudly on so many frequencies that it is virtually impossible to miss. The signal is repeated often enough for it to be recorded before it ceases. Analysis of the signal indicates that one small section is a description of our Solar System but with a peculiar emphasis on the fifth planet, Jupiter.

Some inquisitive little radio telescope is going to point at the giant world to see if there is any response. Needless to say, there will be. A new and obviously artificial radio source will then start beaming out from Jupiter. Closer investigation will show that the source is actually in orbit about that world. We will then try contacting the artificial source by radio and, of course, we will be quite successful. The source will be far enough away not to be regarded as a threat, so mankind will ponder what to do next.[5]

The chances are that we would decide to send an unmanned fleet of space probes out to observe more closely. In any event the ETI in Jupiter orbit will send us a clear unequivocal invitation to visit them at Jupiter. They will not indicate any intention of coming across the asteroid belt to the inner planets. They may even state specifically that they have no desire to do this, or cannot do it because their ship is crippled or some other excuse. Principally they will be staying at a safe distance. 'Safe' for us, that is, in terms of human psychology.

And, of course, eventually we will go in person—sooner rather than later, I think. What will be miraculous will be the rapidity with which the ETIs pick up the major spoken languages of the human race. I would even go so far as to suggest that they insist that the expeditionary crew consist of particular types of people whom they know will be especially susceptible to bonding at physical encounter. Of course they will know us well enough to disguise this motive.

Imagine a group of sex-starved young bachelor males boarding an ETI ship and finding themselves in the company of the most physically delightful and exciting beings they could imagine. Bonding individual human personalities to individual ETI fabricated beings would then take place. The men would be told a lot about the true nature of the Universe and introduced into the mysteries of the Galactic Culture through an initiation ceremony which bound them to the fraternity of Galactic Minds. Some hypnotic implants could be used here, but the chances are that they would be entirely unnecessary. This secret-society bonding appeals to something basic in the social equipment of male humans, and our ETIs would certainly not overlook something as useful as that.

As the decades passed the ETIs would be accepted and their 'crippled' ship in Jupiter orbit regarded as an embassy of an alien power within the Solar System. Then slowly, gradually and gently the ETIs would be able to introduce the truth about themselves to the human race in a very controlled environment. At this point the scenario must end. There is no way of knowing what the ETI motives are likely to be. We do not have the opportunity of closely analysing their biogrammar.

One thing is fairly clear though. They will introduce us to a new and almost certainly uncomfortable perspective on what it is to be human and evolving in a Galaxy brimming with myriad different forms of mental life.

5

A Meeting with Tomorrow

So few people find evolution a magical and exciting subject. This has often left me quietly mystified. I wonder whether or not it is I who have a peculiar perspective on the phenomenon of life ever-changing. Most people regard evolution in the same light as they regard the path that leads up to their front door: they see it as something external to where they live, static and not really all that relevant to what goes on on the other side of that door. But to me it is a roller-coaster ride carrying us all from the black silences of prehistory into a roaring future. There is something heady and perpetually relevant about it. Teilhard de Chardin put it like this:

> ... it is a general postulate to which all theories, all hypotheses, all systems must henceforward bow and which they must satisfy in order to be thinkable and true. Evolution is a light which illuminates all facts, a trajectory which all lines of thought must follow ... [1]

And yet so many of us consider Man to be the supreme point of evolution. With us it has achieved a self-conscious being. From this point evolution leaves off in the biological sense and cultural evolution takes over.

Homo has been around for better than a million years and *Homo sapiens* for over a quarter of that time. Certainly it can be pointed out that there have been no major changes in the human anatomy in the past fifty thousand or even hundred thousand years. It would be foolish indeed to imagine that nature has therefore 'switched evolution off', so to speak. The shark has been around for around fifty million years in more or less its current form and some of the social insects have been around for considerably longer. Are we thus to imagine that those creatures share some kind of evolutionary peak with us? Of course not. Those animals have simply adapted to a highly efficient form, their optimum evolutionary status. Whenever major mutations appeared they were generally less adept at surviving than the original, and accordingly died out.

Has this happened with Man? There are those who claim that we may evolve into particular types of beings. Generally the argument runs along the lines that, as we sow, so shall we reap. The advocates of what I shall call natural evolution tend to paint similar portraits of the beings we are to become. The natural evolutionists argue that, as mental activity is one of the main factors which has put Man in a position of dominance over his fellow creatures, natural selection will favour the bigger and better brains. As such appendages as arms and legs become less and less important, due to the facts that we will have artificial transport to zoom us around everywhere and machines to perform all of our physical work for us, our bodies and limbs will become thin and frail. In fact, we will look like nothing so much as giant babies. Hair will vanish—as it almost has already. Our heads will be enormous. And physically we will be virtually helpless without the aid of our machines.

I was very pleased to hear that this rather disgusting description of the *Homo superior* is in truth rather an unlikely one. Apparently the width of the female hip bones is critical where it comes to the size of the child's head. That width has already spread as far as it realistically can. To spread further would entail a change in the stance of the female of our species. She would have to go around on all fours. Appealing as this may seem to revenant anti-feminist males, this is unlikely to meet with universal appeal.

To compensate for this fact, the human brain has probably become increasingly complex and convoluted over the past half million or so years. This process of wrinkling up the brain's 'skin' to increase its surface area also seems to have gone just about as far as it can. So, where do we go from here?

Science fiction has provided a number of suggestions. Most popular amongst these is that of acquiring extrasensory powers. These may very well be just around the corner. Soviet research in the field of ESP is extensive, and the whole field appears to be gaining some measure of respectability against great odds even in the United States. Of course, a large section of the scientific community is whole-heartedly in opposition to anything even smacking of the paranormal.[2] This is healthy. If the paranormal is scientifically valid it will come through eventually; if not we will be saved from our own attempts at wish-fulfilment and a grand variety of fraudsters.

If future human evolution lies with the paranormal there is currently very little that can be said about it. Certainly, if extrasensory powers do become widespread, the impact upon social life will be at least as devastating as television.

Then let us assume that Man has reached very rapidly indeed the condition of the shark; say that we have reached a dead end so far as potential for further physical evolution is concerned. We are then left with the cultural evolution which I mentioned earlier. Where does it go from here?

Cultural evolution has so far taken us from living in tribal or pack units in or around caves to the heights of sophisticated social life today. These changes have transformed our lifestyles dramatically, particularly over the last twenty thousand years. The great ice masses lost their grip on the planet and Man gained his.

There is no reason to suspect that the changes which cultural evolution will subject us to over the coming twenty millennia will be any less profound than those of the last; in fact, should we survive in any form, they are more likely to be even more fundamental in their impact. They will change the nature of our social organization quite as thoroughly as they did before. But, more important, they will also change Man—and more drastically in his form and nature than in any change the world has seen in the seventy-five million years or so since the Age of Mammals beagn.

We are now in a position to see physical and cultural evolution merging at that point on the horizon towards which we are rushing.[3] The amalgam into which they are about to be synthesized bears closely on our attempts to understand the problems posed by ETI encounter. If, for example, I am wrong about ETIs being basically different from one another in the format of mental life then they will probably be somewhat similar to one another and to ourselves.

We will in effect be, in that case, encountering the equivalent of the human race at a different point in its development. The Galactic timescale is so vast that even the closest encounter is likely to be with a race ten or twenty thousand years ahead of us; in effect, this amounts to us meeting our descendants of many millennia hence. So at this point it is necessary to investigate the probable future of mankind and particularly that point on the horizon where our evolution reaches a critical stage.

Evolution means change. Most people do not respond favourably to the idea of change at all. If they do, it has to be 'change for the better'; and this statement tends to vary in its definition from individual to individual. Large-scale social manipulations for change fall into the realms of professional party politics on the national and international scale. What is a 'change for the better' *here* is *there* a matter of furious debate, frequent exchanges of ill-feeling, and all too often war. The latter condition itself can hardly be regarded as a 'change for the better' in any but the most extreme cases. But like it or no we have to agree with Heraclitus when he said that nothing is permanent but change: evolution, cultural and/or physical, never stops.

We may not like what we see ahead, the possible routes for the evolving Man, but we can only change them; we cannot isolate ourselves from them. Man, the human being in physical, intellectual, social, emotional and spiritual terms, is going to change. There is no way that this hard fact of existence can be avoided. We may dream of remaining more or less as we are, with more sophisticated technology perhaps and more rational systems of law and government. But if this dream has as its basis mankind remaining fundamentally the same as it is today then truly it is a dream.

Before we can approach that crucial point on the horizon in this investigation it is important to examine the dream and to dispel it.

Most science-fiction stories set in the centuries ahead tend to perpetuate the dream because it is much easier to communicate with people in our own society by setting the novel or short story in some recognizable variant of that society. For this reason I am going to take as an example the society which Larry Niven and Jerry Pournelle created for their joint work, *The Mote in God's Eye*.[4] (This is not to be a criticism of the book, which personally I found immensely enjoyable; this is simply an examination of one variant of the dream.)

By the year 3017, one thousand and thirty-nine years ahead of the time at which I am writing this chapter, Man is spread across the stars in a vast Empire and, like most empires, this one has troubles with dissidents. The opening pages concern themselves with a band of rebels being quelled by Imperial troops on a planet known as New Chicago.

One thousand and thirty-nine years in our past the Arabs were having even greater difficulty holding their empire together, losing Madrid to Leon that year (939). In fact, the whole future chronology seems to relate quite closely to the past two thousand years of European history. There are colonizations, coronations, empires, wars and even society crumbling back into the Dark Ages only to emerge triumphant again. Bringing the future properly and democratically up to date, the Emperor appears to be a constitutional sovereign inasmuch as he has his Viceroys dotted about the Galaxy; these in their turn appoint governors to the various planets. He has also an Imperial Parliament to which elected members may be sent from the various worlds, all of which have their own democratically elected assemblies.

There are many other variants of this dream: some see in to-morrow the historically 'inevitable' emergence of a communist Utopia, others see it as the territory of supercorporations where nations are but token institutions. Whatever the dream, it always has its roots firmly in the present or in the past. There seems no other way of looking at the future constructively. Unfortunately the dream takes that which is already established and modifies it only marginally and usually superficially when placing it in a future context.

We tend to laugh at portrayals of the twentieth century seen through the eyes of the past. They seem ludicrous because so much has taken place in the interim which the previous writers have missed. For example, there is the anonymous book published in 1763 entitled *The Reign of George VI*.[5] In this the twentieth-century King of England is victorious at the end of the European Wars. This does sound vaguely like the stuff of prophecy until we realize that this King George is an idealized eighteenth-century-style monarch who leads his troops into battle on horseback with unsheathed sword. Ships of the line blaze away at frigates in set-piece naval battles and the most revolutionary weapon appears to be the fireship.

Of course, it can be argued that SF in the mould of *The Mote in God's Eye* incorporates the concept of great technological advance-ment (the Alderson Drive which gives Man the stars). In fact, a great deal of SF, including this major work, idealizes the techno-logical exploitation of scientific discovery much in the same way as *George VI* idealizes the concept of kingship. Conflict arises not from the idealized subject matter but among the personalities involved with it.

Here we have the key to the dream: it lies in a kind of perfec-tionism. By this I mean the 'perfected' oppressive State which Orwell described in *Nineteen Eighty-four* and the 'perfected' Advertising Agency world, in which humans have the status of consumer units, in Pohl and Kornbluth's *The Space Merchants*.[6] In perfectionism, single aspects of present-day society dominate the future in the way their authors see them influencing the present.

History does not seem to operate like this, however. It quite obstinately refuses to be predictable, reading more like a catalogue of surprises than an even semi-logical progression of events. Most writers who set a work in the future do not expect their readers to regard it as a predictive statement. The best are often cautioning in their tone—not so much dreams as nightmares. Many, unfortunately, postulate a future intended as neither utopian nor anti-utopian but as arguably probable. The failing here is that few have an evolutionary perspective, even socially.

For example, the sophistication of political thought over the past millennium has been a major factor in the shaping of political systems and groupings in the world today. To assume that there will be no innovations in this area is surely naïve.

So, when we are postulating a future for Man, we must postulate on the basis of what the major innovative changes are likely to be rather than on what aspects of life today are likely to dominate life tomorrow. The major innovative changes are those introduced by new and valid concepts—so how can we postulate, as these are presently unknown?

What I am going to do is speculate that changes will take place in areas where they are almost certain to occur, and areas, at that, which are crucial to the evolution of human beings both as social creatures and as a species. If these are in any way even near accurate then the Galaxy may very well be fairly well penetrated by our descendants in the year 3017, but by that time *Homo sapiens* will almost entirely have vanished.

Specifically, the area from which I see the evolutionary leap emerging is cybernetics: 'cybernetics' was defined by Norbert Wiener[7] as the science of control and communication in the animal and the machine. Essentially, the areas which will be involved in the transformation of Man are genetic engineering and intelligence simulation in machines. No one would be more surprised than myself if these proved to be the only areas of technology within which breakthroughs proved to have dramatic social consequences. I would suggest that we will see some startling steps in material science ('smart materials' programmed to alter physical properties and shape in various circumstances), data storage ('pocket'-sized libraries), and in learning technology (drugs which enhance both long- and short-term memory).

But, to return to cybernetics, I would say that it is here that the most obvious evolutionary impacts can and will be made. Control of genetic information is crucial where a new species is to be created to any preconceived design specifications. As I am concerned with the creation of a new species (or several thousand new species), a degree of sophistication which we do not yet possess in biological engineering is a prerequisite. But much more important than this is the actual willingness to use that sophistication on the DNA, the genetic code-stuff of human beings. This, after all, is the ultimate in eugenics.[8]

Eugenics raises a number of fairly major moral and political problems (as well as hackles). The very fact that these will have to be wrestled with means that, when mankind comes to use his genetic wizardry on himself, the moral and political climate will be changing accordingly. This illustrates that technological progress is not isolated from social and political progress.

Perhaps there are those who doubt that the human race will be changed by eugenics, by a genetic policy. Is there a good reason why eugenics is inevitable? Well, perhaps it is not. I can imagine that, with societies run according to the low-scale alternative technologies advocated by some followers of the Club of Rome, humanity would be fairly safe from the horrors of the ravaged environment, the bomb and the 'terminal laboratory experiment'. Unfortunately, the economic changes which this type of society requires would in my opinion create traumas of a completely unacceptable level in politics, military circles, and the financial establishment. It would mean the end of the growth-oriented society. I do not believe that this is yet in sight.

To be basic about it, eugenics, unlike limited growth economies, does not affect the voters' bank balances adversely. People generally prefer to anguishedly clutch the brow rather than the wallet and most politicians tend to respond to this accordingly. Consequently I tend to think that we will see eugenic legislation in the field of human genetic engineering quite some time before deliberate attempts are made to profoundly change the growth orientation of society.

Apart from the fact that engineering eugenics could eradicate such things as sickle-cell anaemia, mongolism and spina bifida without resorting to restriction of the breeding of couples linked to these, or even sterilization, there is the commercial aspect, which could contribute to the growth-society philosophy. 'Ensure your child's high IQ for three thousand dollars': specific highly prized characteristics could be enhanced in the genetic material.

Already there have been a number of experiments using the genetic matter of human beings; for example, the creation of hybrid man-plant organisms has been successfully accomplished. The really important step comes when we attempt to interfere not with the human body but with the mind. I believe that the day will come when we deliberately set about the process of endowing our progeny with the intelligence levels of geniuses simply so that they will be on a level with their peers. The general idea of brain enhancement will then be a commonplace.

Already there are drugs which combat the mental effects of senility by slowing them and in some cases even reversing them. It is a logical step from this to drugs which aid the brain's functions in other ways: memory, concentration, imagination, emotional control and so on. With this psychologically acceptable, the concept of 'plugging in' the brain to a computer becomes palatable. We are then into the area of computer-aided intelligence.

But the computers themselves, by this point, will have moved on from the current generation of machines to a state of the art where the privacy of the individual will be seriously threatened. This is already apparent today. Grosch, the Cassandra figure of computing, has pointed out that, with the fourth generation of these machines emerging, we are threatened with 'invasion of privacy, domination of law processes by FBI type agencies'.[9] Rather gloomily, but realistically in this context, Grosch's Third Law states that 'things will get worse without limit'.

Indeed they will, and ironically the best defence against invasion of privacy will be access to your own high-performance computer. It will have to be one which can be isolated from the growing networks of computers and inaccessible to the various methods of electronic burglary. This may simply mean that all the data and operational programs which are relevant to you are restricted to your usage by some complex codes and passwords—rather like the electronic equivalent of a safety deposit box.

The machines themselves will be becoming much 'smarter'. This will be due to the increasingly efficient design of not so much the machines as of the software, the programs. Whether or not genuine artificial intelligence is created in this area, a certain amount of machine intelligence will be used. This will take the form of programs designed to mimic intelligent actions in certain restricted situations.

Already computer programs are quite heavily employed in the

business of designing other computer programs; it seems that the logical extension of this would be the machine system which was capable not only of designing itself but a more efficient version of itself, a 'smarter' version. Obviously, if this new machine is built it will be even more capable of this feat than its predecessor . . . and we find ourselves with an exponential rise in the performance characteristics of the machines.[10]

This does not mean that these machines will eventually turn into intelligent beings. They will still be restricted inasmuch as the data base upon which they will be operating will, we assume, be that of the original machine. Within that restricted system they will become superior to the capabilities of Man: becoming superior to Man himself would mean that the system would have to have as its data base the information present in the brain of a new-born child plus a design programme to enhance the design of the system as a whole.

Most computer people to whom I have talked about this matter, including those whose work involves looking at where computer design is likely to go, regard this as totally impossible. Being an SF writer I tend to dismiss their highly qualified opinions with my unqualified and mindlessly confident one that superior artificial intelligence is feasible. Where I do to some extent agree with the computer specialists is that I do not see the 'god-like machine' arising from this area. Not on its own, anyway.

We are at the position where there is a direct interface between the human brain and the computer. I would like to predict that the feedback from this combination will be so staggering that more research will go into the enhancing of this relationship than anything related to artificial intelligence.[11] This would bring about a position where the minds of geniuses by the million are being enhanced, artificially boosted by super-computers capable of handling masses of information and flickering it selectively into those brains as it is required. No need for libraries; all the information relevant to a subject is there and being referred to as the genius simply thinks about it. In this situation one psychopath might dream up overnight half a dozen feasible plans to destroy civilization and wipe out the human race several times over.

Our generation of geniuses will realize this long before the possibility ever becomes real. There will have to be two basic safeguards. Firstly, all machines will have to be limited; none will be able to assist such thinking and, once it has identified a psycho-

78 A MEETING WITH TOMORROW

pathic thought pattern, a machine will automatically alert Grosch's FBI type agency, Orwell's Thought Police. The second aspect of safety will be eugenic. Man will be made biologically incapable of killing his fellow man in almost every situation. This limiter may be written into the DNA itself, if possible. The release mechanisms which would allow such an action will be contained jointly in Man and in his symbiotic mechanical aids.

With the emergence of the man-machine hybrid already performing surgery on the nature of human nature we see the curtain falling on *Homo sapiens*.

Whether or not the 'Hybrids', as I shall call them, wish to maintain a human form they will almost certainly go about the process of farming better minds for themselves. The brains will be grown as clones from originally human genetic material. They will probably be designed around the Hybrid needs and, at present, it is impossible to say what these are. From our own point in time we might speculate that the surface area of the neo-cortex might be increased (folded more complexly) and that the newer parts of the brain be linked with the older parts in a fashion whereby the newer part was always in control of the relationship.

Alternatively, the actual organic material—nerve tissue, in other words—may be forsaken for something totally different, silicon crystal perhaps. Instead of designing a brain in this material the relationships between the various brain cells which are expressed through the synaptic connections would be copied exactly with the much smaller interstitials between the crystals taking the place of the synapses.

What would such Hybrids be like? To begin with, they would be virtually immortal. Their intelligence levels would be meaningless, in real terms, to us. They would be telepathic to all intents and purposes as they would be able not only to transmit thoughts to one another by electromagnetic impulses but to conjoin into 'think tanks' the likes of which it is pointless to speculate upon.[12] They would not be limited to our five senses; they would be able to smell colours, taste heat, feel aromas and so on in environments where *Homo sapiens* would have blundered about swathed in protective clothing. Their emotional lives would have a potential richness beyond all imagining.

What would they look like? Not like us. And who can really say? Perhaps like great crystalline blocks containing a million minds or more and speeding out towards the star clouds. Perhaps they would have one or more human-type forms which they could occupy by remote control if they so desired.

And, finally, when?

Guess hazarding is always risky in the prediction business. It might be possible to produce the first Hybrid within the next one hundred years, perhaps a good deal sooner. On the other hand my predictions could prove as useless a guide to the future* as any other exercise in speculation based on personal belief.

But, if I am right and if our encounter is with creatures who are representatives of what amounts to our future not a century ahead but some *two hundred centuries hence.* . . .

*See Appendix 4.

6

Saucers and Dishes

There are many books which act as excellent outlines to CETI's history and current thinking.[1] It is not my intention to give a detailed resumé of what you can find in these texts but it would be remiss of me if I did not include a rough sketch of what has taken place in the field during the past three decades.

Most people seem to be under the impression that there has been little, if any, serious scientific work done in the entire area. This is probably more so the case for those who live in English-speaking countries; they will be well acquainted with the steady stream of derision which has poured from the establishment—both scientific and bureaucratic—on the topic of UFOs. However, UFOs and CETI should not be confused. The areas of study have quite distinct roots.

Reports of strange 'foo-fighters' were made mostly by military pilots in the second half of the nineteen-forties. These became a craze or cult—call it what you will. There has been a lot of 'evidence' but none that leads to any conclusion universally accepted regarding the nature of these weird happenings.

There is no connection between that and the emergence of the idea of CETI. Public enthusiasm for UFOs, which frequently verged on the hysterical during the 'fifties, tended to make scientists keep their mouths shut about private speculations on intelligent life existing elsewhere.

CETI is a dangerous area to tread around, which is one of the reasons that it attracts certain types of people. Furthermore, there are more hazards than simply being tarred with the 'UFO-freak' brush. Nature and colleagues alike can make the CETI-interested scientist feel idiotic and the object of laughter from others. There is an old Arabic saying: 'The dogs bark but still the caravan moves on.' This is very much the attitude which the scientist has to adopt in such cases.

As I said above, speculation on the subject by scientists was very low-key during the early and mid-nineteen-fifties. If anyone talked about CETI they did so in private to equally interested friends. One such was a physicist by the name of Giuseppe Cocconi, who had discussed the possibilities with Philip Morrison, another physicist. Excited about the possibilities of ETI broadcasting radio messages, Cocconi wrote a letter to Dr (later Sir) Bernard Lovell on 29 June 1959. He presented in this his case for utilizing the new radio telescope at Jodrell Bank in a search for radio evidence of ETI. Lovell, the observatory's director, replied briefly and disappointingly, according to Cocconi. In fact, Lovell found the suggestion 'frivolous'.

Not to be deterred, Cocconi and Morrison wrote a paper in August that year for *Nature*, Britain's august scientific journal. In this they speculated that radio broadcasts might be found in the region of the hydrogen emission line in the radio spectrum at 21cm (1420 MHz) which might act as an electromagnetic landmark for such activities.

One of the world's most eminent scientific publications, *Nature* was an excellent choice. In the past it had presented a good number of equally sensational and imaginative concepts within its pages. This one was to prove no exception. The proposal appeared in the issue for 19 September that year.

At last it was possible to talk openly and respectably about the subject, if at some distance from the ordinary public and safe from the ridicule of the press. One of the startling facts to emerge was that a search along the lines suggested by the two physicists had been planned in virtual secrecy at the National Radio Astronomy Observatory at West Bank, West Virginia.

Frank Drake, an astronomer working there and a graduate of Cornell, had proposed a similar scheme to Lloyd Berkner, the Acting Director of the observatory. The older man expected that such a scheme would be scoffed at but he evidently believed this particular caravan would move on because he encouraged Drake. He even went so far as to bring the subject of Drake's proposals up with the man who was to be appointed as the full time head of the observatory, Otto Struve.

The project was initiated in April 1959. The following month Struve took over the full-time job of director and was largely responsible for running the project. Drake called it Project Ozma after the Princess Ozma of the exotic land of Oz conjured up for children by Lyman Frank Baum.

In the spring of 1960 the search began. There was a great deal of mixed feelings within the scientific community, and no one was particularly surprised when the two stars which had been repeatedly scanned, Tau Ceti at almost twelve light years distance, and 61 Cygni at just over eleven, rendered no significant results.

There came the expected hoots of derision and noisy raspberry blowing but nevertheless CETI had arrived. Even Sir Bernard Lovell mellowed. By 1962 he felt sympathetic to 'an idea which only a few years ago would have seemed rather far-fetched'.

There are three possible areas of search for CETI (or SETI—search for extraterrestrial intelligence).

The first is for astronomers to make a search for radio signals. I put this first because it is the one receiving the most attention by members of the scientific community, not because I believe it to be the best (I most definitely do not). There are various different types of communication of this sort which we may intercept. We may tune in to a beacon broadcasting generally into the Galaxy or specifically at us. We may also pick up a message being transmitted between two ETIs already in mutual communication. Lastly, we may 'overhear' the domestic radio traffic of another planet much in the same way that they would be able to 'overhear' our own. Remember that, by the year 2000AD, mankind's electromagnetic squawking will be 'audible' over seventy light years in any direction in which you care to point, a volume of some one and a half million cubic light years. And, if we can hear them, they can hear us.

Radio signalling should really be termed EMG (electromagnetic) signalling. There is a good deal of discussion concerning other areas within the electromagnetic spectrum which might be useful for interstellar contact. Lasers have been proposed; not merely optical lasers either. There has been discussion on the use of X-rays, gamma rays, neutrinos, etc. The basic beliefs on strategy, however, remain the same: interaction between ETIs in different stellar systems can take place only by means of signals.

The two other avenues open to SETI are not given substantial credence by most workers in the field. This is probably because they verge on the sensational rather than because of any arguments which have been aligned against them.

The most sensational of these is, of course, the possibility of an ETI presence here on Earth at the moment or at some time in the past. As far as the present is concerned, UFOs are generally what the public think of when CETI is mentioned, spacecraft crewed by verdant antennaed dwarfs. Close on the diminutive heels of these cute little characters come the ancient and semi-divine astronauts. How sad it is that public thought on the subject of CETI is limited to these two.

No one knows, or is sane and admits to knowing, of a way to prove what all UFOs really are. For this reason, if none other, we should not dismiss completely out of hand the possibility that there may be an ETI explanation for some reports. There may also be psychical, religious, occult, or even perfectly natural terrestrial explanations. We cannot prove anything concerning UFOs—that is, truly *unidentified* flying objects.

What we can do is take a look at the hypothesis that they really are ETI spacecraft and see how well it stands up. From reports we can estimate how rapidly they accelerate and how they move. We know of no propulsion system capable of doing this. A direct consequence of the movement pattern entails g forces which would cripple or kill the average human being travelling inside and tear apart the structures of the hardiest flying machines which we have built. They are not built or propelled according to the principles which we apply to spacecraft and they are not peopled by beings physically like us unless there is some field force, currently inconceivable to us, which effectively isolates the crew from the crushing acceleration.

All right, for the sake of argument let us accept that some few are spaceships. What does this tell us about the Universe, particularly the Galaxy in which we lie? An American physicist, Hong-Yee Chiu, has worked out that the mass required to build even a small number of UFOs arriving each year at every star in the Galaxy means that by now some two billion stars, or about one percent of the Galaxy as a whole, would have been mined.[2] The problem is that, even if this massive industrial operation is under way, we have no way of detecting it. Perhaps stars which are being 'mined' show spectral anomalies, but how are we to distinguish the spectral anomalies which are artificially created from those which are created by natural processes of which we are not yet aware? Even if UFOs are being mass-produced like spacefaring widgets, we cannot say so from our present investigations.

Placing the UFO–spaceship hypothesis in the 'things we have yet understand category' does no good whatsoever. This is the *ad hoc* hypothesis category where we again find the psychic, the religious and the occult (UFOlogists might disagree).

Back to square one. In time all *ad hoc* hypotheses may become subject to proper scientific scrutiny, but that does not help us here and now.

So what about the ancient astronauts? The works of Erich von Däniken do not comprise my favourite reading but there is no point in my criticizing his works as the job has been done expertly by Ronald Story. Story's surgically incisive investigation, *The Space-Gods Revealed*, dismembers von Däniken's theses with a flourish which I imagine has not been seen since Rome banned gladiatorial combat from the Coliseum.

The potential investigator of SETI-archaeology, searching the past for evidence of ETI visitation, must be very careful. He is liable to be tarred with the von Däniken brush. For some time to come those who are interested in this topic may well adopt a low profile, as did CETI during the 'fifties.

An artefact which evidently was the product of an advanced technology and was discovered in an archaeological dig among material from a time when, to our knowledge, no such technology could have existed would open the possibility that ETIs visited Earth in the past. It would also open the possibility that there was once a very advanced terrestrial civilization of which we have no record. Several such artefacts might render a better picture of the technological capabilities of the originators, but a good deal of material would have to be amassed before either the ETI or the purely terrestrial model was ruled out.[3] A richer approach should be the search for that which is of indisputable origin. A large installation upon the Earth now covered by millennia of surface activity might be worth looking for. A databank left deliberately for us would be even more interesting. The problem is, where might the departing ETI decide to place the databank?

We know that if they were here and did leave one, they intended that it should be found. Thus we can assume that they would have taken steps to make sure that in a given set of circumstances we would find it. Following from this we can say that there are three possibilities relating to its discovery.

First of all, the databank may have been intended for us to have access to at an earlier period in our history. If so, we can imagine that the given set of circumstances changed in a manner unforeseen by the ETIs after their departure. This could have been an earthquake, volcanic eruption or other natural catastrophe.

Secondly, we may not have the ability to appreciate the significance of the set of circumstances or recognize the databank for what it is. If this is a matter of an evolutionary quirk it might be an extension of the first possibility, in that either they anticipated another animal would evolve mentally faster than our ancestors did or that we evolved in a different manner to that which they expected.

The last possibility is that before now (and perhaps not even now) the set of circumstances has not been complete. For example, a standard degree of technology may be a prerequisite for our appreciation of the databank indicator. SETI-archaeology would, naturally, be another prerequisite in many such cases, although not all.

Only a civilization with knowledge of radioactivity would appreciate the singularity of the natural atomic reactor in French Equatorial Africa.[4] A suitably insulated databank might be placed inside this with the knowledge that only a human society with a fairly highly developed scientific capability would investigate this phenomenon.

If the departing extraterrestrials were interested particularly in those members of our society who would be pursuing SETI they might have chosen an unmistakably extraterrestrial reference point on the surface of the Earth. The centre of South Africa's Vredevoort Ring meteorite impact crater could be a candidate, or perhaps one of Mauritania's larger Richat craters, should these prove to be meteoritic in origin.

In the event of our morality being a particularly crucial characteristic which they would have wished to include in their key set of circumstances things grow much more difficult. They might have left the databank concealed in such a way that only a civilization which had passed through our present nuclear adolescence would discover it. This might entail placing it off the Earth, a possibility which we will explore in a moment.

They might have wished to leave it to a society which is conservation-oriented, non-xenophobic and interested in communication with other AMLs. Such a race would be ideal to move among the stars doing minimum damage to the Galactic environment or to the beings which inhabit it. To accomplish this, the databank should ideally be embedded in an aspect of our environment threatened by our destructive impulses, ravaging greed and the urge to eliminate any other AML. Could the databank be genetically buried in the enormous complexities of the cetacean brain, say? Could this even be the reason why these animals have such large and complex brains? Are they in fact aquatic librarians with enormous store-houses of knowledge locked away within the recesses of their non-human minds?[5]

The final prospect of SETI is that tangible, irrefutable evidence lies beyond the Earth but within the Solar System,[6] a possibility explored in the finest of all English-language SF movies *2001 : A Space Odyssey*. Artefacts on the Moon would not be subject to the ravages of surface activity on Earth. Even simple oxidation would be absent and, in the case of a buried object, smaller meteor impacts would be harmless: the electromagnetic sunstorm which the lunar surface is subjected to would be at a safe distance above and the temperature flux would be suitably low.

Where should we look?

Again, the most prominent indicator of an extralunar connection might be a useful starting point. In this case it proves to be the quite startling Mare Orientale, a vast complex crater structure unlike anything else on the Moon. It is further significant in that it cannot be appreciated when viewed from Earth as it is always seen edge on. Perhaps some time in the opening decades of the next century a team of selenologists might uncover something rather unexpected right in the centre of this remarkable feature.

Alternatively, the databank may have been placed in an orbit about some body in the Solar System. Perhaps it follows a cometary path which brings it close to the orbit of Earth every few decades. It may have visited us during the Second World War, perhaps, and we were too preoccupied to heed its attempts at communication. Such a databank would probably stay closed down for most of its orbit, powering up only as it approached and monitored us and perhaps attempted to make its presence known. On the swing away from the Sun it would close down, performing as few maintenance activities as were needed through the preserving cold until it again approached the Sun and Earth briefly. To ensure that the other planets did not distort its orbit too greatly the databank would possibly be placed in an orbit inclined at ninety degrees to the plane of the ecliptic, the great invisible disc with the Sun at its centre and the orbits of most planets inscribed upon its surface.

This leads us straight on to our third possibility, the proposal that there may prove to be a probe from another star system orbiting our own Sun and perhaps even in orbit around the Earth. This was first put forward as an idea by Professor Ronald N. Bracewell of

Stanford University in an article which he wrote for *Nature* in 1960.[7]

We send out numerous robot explorers to the planets, so it seems fairly reasonable to suppose that another star system which evolved an advanced form of mental life with a desire to reach across interstellar space might dispatch a similar device towards us. If it did not simply fly though our Solar System but took up residence here there is a possibility that we might make contact with it. Even if it has been in space somewhere fairly nearby for hundreds or perhaps even thousands of years it might still be functional. Wear and tear from the environment out there is in some ways rougher, due to radiation exposure and temperature extremes, than might be expected on Earth. The compensation comes in the form of missing Earth's atmospheric effects, such as wind, rain, oxidation and other corrosive activities.

This is quite exciting because, even if the spaceprobe is inactive, it is at least something possibly resident in our system. It gives us the possibility of something substantial and fairly permanent to search for instead of the fleeting radio wave. One probe discovered would probably yield more information far more rapidly than an intercepted radio signal, particularly at first and certainly if the radio signal is not a beacon but either domestic radio traffic or a communication from one ETI to another.

The general criticisms of the spaceprobe hypothesis are founded on efficiency. First of all, it is a very expensive method of going about CETI. The costs involved are enormous in comparison with the transmission of radio information. An ETI employing spaceprobes for CETI purposes similar to our robot explorers would probably have a very different appreciation of priorities involved in the subject than most CETI thinkers here on Earth. Cost would not be one of the prime considerations.

Another criticism of the spaceprobe hypothesis rests on the distances involved in flights between two ETIs. Here there is the problem that, by the time the spacecraft reaches its target star system, the home system ETI may have passed beyond its lifespan. Considering that interstellar flights may entail coasting through space for decades, centuries or even millennia, this is possible.

Alternatively, an ETI which realized that it was soon to perish might send out a number of such spaceprobes to likely nearby star systems. These would effectively be time-capsules or databanks which might indicate to an emerging intelligence the problems which lay ahead, the strategies which the ETI employed and why it ultimately failed.

The model of the development of Advanced Mental Life in the Galaxy which makes it somewhat more fruitful to search for such probes is the one where there are very few active ETI intelligences in the Galaxy at any given time. This means that the lifetime of the ETI once it reaches the appropriate level is all important. In this sense the ETI has to be regarded not merely as a being or group of beings but a condition which involves sentience with a long developmental history in terms of knowledge and capabilities. In Man this is best expressed in terms of civilizations. Obviously the knowledge and capabilities of a civilization greatly exceed those of any individual within it. Accordingly we should require a similar level of complexity in the ETI.

How long an AML remains at the level of being able to indulge in CETI and wishing to do so is a critical factor in assessing the number of such AMLs. This number determines the average distance separating them and thus has a direct bearing on any exchange of signals. If there are many such ETIs, we can expect there to be a fairly active radio discourse taking place; but, if there are few, there may be very little if indeed any communication of this sort at all.

In the event of there being few CETI-oriented lifeforms, they may not survive sufficiently long to set up a dialogue across the vast reaches of interstellar space. Bracewell has estimated that, should we receive a signal from an ETI some one and a quarter thousand light years away our reply should reach it, if it were sent immediately, just at the point when that 'civilization' was either perishing or moving out of its CETI phase. If this is common, some ETIs may well place robot databanks in other likely star systems. However, it may prove to be a better and cheaper strategy to place automatic addressable beacons marking the passing on of such mental lifeforms broadcasting from their home system.

The construction of spaceprobes purely for the purposes of seeking out and establishing contact with ETIs is an approach which we humans would be most unlikely to adopt. This does not mean to say that an ETI would eschew it but it does mean that psychologically we are likely to be biased against investigating the chances of such probes being resident in the Solar System.

An extension of this is the starship peopled with ETIs entering the Solar System. This is a circumstance which has been given a great deal of coverage by science-fiction writers but virtually none by serious students of CETI. Yet, today, we are beginning seriously to consider the possibilities of interstellar flight and its problems.[8]

There is no way of logically demonstrating that we are more likely to hear a radio signal than be visited by a spaceship. The choices are based on what we know of the technology of the one AML with which we are familiar—ourselves. The drawback here lies in the fact that we have a disquieting historical perspective to bear in mind. If the scientists of a century past had been avid CETI fanatics they would have searched their known Universe vainly for signs of intelligent life elsewhere. They would not have considered radio because they knew nothing of it. Philip Morrison points out that we may be in the same position and that the ideal method of interstellar communication is by 'Q' waves, as yet undiscovered by terrestrial science.

This has not impaired the intense discussions on the use of radio for CETI. Indeed, over the past two decades, this facet of the subject has dominated the whole to such an extent that it has become something of an establishment attitude. It is now those who do not subscribe to the interstellar radio-signal hypothesis who must smile as the custard pies arrive.

Perhaps this is only fair, as the radio buffs have been squarely buffeted with these metaphorical missiles for some time now, and quite soundly on certain occasions. The Soviet Academy of Sciences, in particular, has had a rather unfortunate history in this respect. On three occasions they have been made to look slightly foolish. The first was the affair of CTA-102, when it was supposed that the Russians had discovered a super-civilization but this proved not to be the case. The radio signal which they were studying proved to be a natural radio source.

CTA-102 (and another source, CTA-21) had been the subject of some conjecture by the Soviet researcher Nikolai Kardashev. In 1964 he had postulated that these might be distant ETI civilizations because their radio spectra appeared quite singular in respect to other known radio sources. The following year came the sensational discovery that CTA-102 fluctuated smoothly in 100-day cycles. Although the scientists were interested it was *Tass*, the news agency, which decided that this was indeed an artificial beacon transmitting across the Galaxy. Pressed by newsmen boggle-eyed at the prospect of handling the story of the century, the embarrassed scientists admitted that, although it was premature to say anything firm, it was at least possible that this could be another civilization.

The custard pie came in the form of an announcement shortly afterwards that CTA-102 had proved to be one of the recently discovered quasars.

More recently, in October 1973, the Radio Institute at Gorki released a communiqué to the effect that it had picked up signals of ETI origin arriving in bursts of between two and ten minutes' duration. Interestingly enough, they seemed to be coming from somewhere within our own Solar System. There was some back-pedalling after further investigation revealed that what had been received was capable of being explained by an artificial satellite in Earth orbit.

The most spectacular suggestion had not to do with radio at all but came from Academician Shklovskii in his book *Universe, Life, Mind*[9] published in Russia for the fifth anniversary of spaceflight. Shklovskii put forward the startling proposition that Phobos, the inner of Mars' two moons, was in fact a gigantic artificial satellite. Sadly, photographs of Phobos now indicate that this is very unlikely, and the information upon which the Soviet scientist based his calculations is now regarded as being suspect.

How then does one avoid the custard pies?

The British example gives us some indication. In December 1967 some very peculiar radio signals were detected by a group of Cambridge astronomers. The signals were very rapid and startlingly regular, just what one might have expected from an ETI beacon. While they were being investigated they were titled LGMs, an abbreviation for Little Green Men. Jocelyn Bell, a research student working at the Cavendish Laboratory with Professor Anthony Hewish, was the first person to come across this phenomenon. After several months of very quiet investigation, during which the possibility that they were looking at evidence of intelligent life far out in space was not easy to dismiss, they came to a conclusion. By this time three other such radio sources had been charted, and close scrutiny showed they demonstrated the basic characteristics of stars. They were in fact natural objects. Today they are generally accepted as rapidly spinning neutron stars: pulsars.

Obviously radio astronomy is a very exciting field to work in. Naturally some of those involved in it tend to frown when they hear talk of their observatories being used for CETI purposes. Their work is so demanding of observatory time, and there is already so little of that available, that they grudge spending it on what they regard as frivolous activity. In a climate such as this it is not surprising that the discoverers of pulsars kept their heads well down for a few months before announcing their incredible find early in 1968.

The search continues. International conferences are now held frequently on the subject. Highly respected scientists the world over are involved in the debates. Perhaps most significantly of all, the tenor of the debates themselves has changed greatly over the past two decades from the theoretical to the practical. It has gone from the basis of trying to make the whole subject appear reasonable and therefore interesting to scientists in general to the examination of engineering considerations entailed in the optimum instruments designed for the purpose.

Again it is in the area of radio communication that most work is being done, and is intended to be done. Since Frank Drake's original Project Ozma there have been Ozma II and a number of other searches of the radio sky. To date, none of these has yielded a significant result. Nevertheless, in virtually every radio observatory in the world today someone has done some hopeful listening for signs of ETI activity.

The Russians are interested in the possibility that there is a high amount of interstellar radio traffic in the vicinity of the Galactic core and are concentrating a good deal of effort on examination of the core regions' characteristics. The giant RATAN 600 instrument used by the Shternberg Institute in the Northern Caucasus is employed in this programme, which aims to scan the sky on as great a range of frequencies as possible. Hopefully a satellite capable of furthering the effectiveness of the Soviet-wide SETI radio-telescope network will be employed.

Kardashev seems at present to be seeking out either a Dyson shell[10] or other evidence of one of his own super-civilizations. These are approximately: Level One, capable of controlling the energy resources of a planet (roughly ourselves); Level Two, capable of utilizing the energies of an entire star system (reducing most of the planets to rubble which orbits the star to form an energy-absorbing shell, and as suggested by Freeman Dyson); Level Three, which uses the energies of a complete galaxy.

NASA, too, is contemplating some exciting developments in the field of SETI. Cyclops is perhaps the most widely discussed system. It consists of more than a thousand steerable radio telescopes each with an antenna more than a hundred metres across. The result would be the equivalent of a single dish well over five kilometres in diameter. The whole system would be integrated and operated by computers allowing it to act virtually automatically. The same computer complex would apply a variety of enhancement procedures to the incoming signals to pick out salient information from the background noise.

There is also an orbital radio telescope under discussion in the USA. This would be a spherical construction assembled in sections brought up by the Space Shuttle. Shielded at all times from the intense radio interference from Earth the antenna would be able to scan the entire heavens in its sweep around the sky. It is proposed to place it in one of the libration points in the orbit of the Moon around the Earth in such a way that it, the Moon and the Earth would form the points of an equilateral triangle—each equidistant from the other.

The most ambitious programme is to set up a Cyclops array on the far side of the Moon. This would be fantastically expensive both to build and to support but it would be entirely shielded from any terrestrial interference by the whole bulk of the Moon.

The trouble is finance.

At the time of writing NASA is still fighting for funds to initiate a reasonable basic SETI programme. There is a little irony in the situation, which relates to the amount of cash they believe they need. The twenty million dollars, spread as two million a year over ten years, is roughly the amount which the film director Steven Spielberg spent on his SF CETI movie *Close Encounters of the Third Kind*. Furthermore, at the time of writing, Spielberg has decided *Encounters II* should have some scenes shot on location and is so sure of his budget that he has booked Shuttle time to do filming in space!

Well, listening is fine, but how can you be sure that there are any signals to hear? How do you know that everyone is not simply adopting the same strategy of merely listening with no attempts at transmission? Well there is one way to be sure. Send out a signal yourself, and you can at least be certain that *that* one is being propagated out across the Universe—and if there is at least one zinging out across the void how many more might not there be of ETI origin?

Today such a little message is on its way from planet Earth to M13, a cluster of stars so distant that the signal will not arrive for

some 24,000 years. It was broadcast from the 300m radio bowl at
Arecibo in Puerto Rico and is made up of 1679 digits. The digits
are a kind of dot-dash system like Morse code except that, unlike
Morse, the message is not made up of letters but is in effect one
gigantic word. The word is translated by straightforward arithmetic.
There is only one way to factor the number 1679; that is to say,
there are only two numbers which when multiplied together will
give 1679 as a product: these are 23 and 73.

It is possible, therefore, to build up a picture from the dots either
23 by 73 or 73 by 23. The latter produces a meaningless jumble but
the former results in a coherent dot-picture which the M13ians
can then attempt to make sense of for some considerable time. It
contains pictorial information about our Solar System, ourselves,
and the broadcasting instrument (which seems to be pointing
upside down in the general direction of our feet, a fact that might
lead the M13ians to deduce that we bounce around on our heads).

We sent it out in 1974 and may thus hope for a reply in the region
of 50,000AD. I would not recommend hanging around specially for
this occasion as you are liable to be disappointed (and after all those
millennia too!). The big let-down results from the fact that the
signal will be difficult to snatch. It is only three minutes or so long
and is being carried on a wavelength which has no particular
significance for SETI (12.6cm was the radar wavelength the
instrument was calibrated for at that particular moment).

Many scientists regard radio investigation of the heavens as the
'uniquely rational way' of going about CETI. This was originally
the insight which Cocconi and Morrison found in the hydrogen
radio-emission wavelength. Since then there has been criticism of
that wavelength and others have been put forward, notably the span
between the hydroxyl wavelength at 18cm and that of hydrogen
(21cm). This has been romantically christened the 'water hole'.[11]

The problem with this rational way is not a matter of whether
there are signals presently sweeping past and through us. The
problem with passive radio listening lies in the thought process,
the very human thought process upon which it is based.

7

Past Contact and the Moving Caravan

A few years ago Duncan Lunan, one of Scotland's few CETI theorists, put forward the thesis that there might be a spaceprobe, similar to the Bracewell type, in orbit about the Earth. What happened then is quite enlightening regarding the way in which various sections of the community reacted to his announcement.

If we wish to understand well how various members of the community are likely to react to the actual encounter should it occur sometime during the coming years or decades we should do more than speculate on the nature of the ETI. We should try to examine some of the attitudes of the throwers of the custard pies.

These are often individuals in positions of respect and responsibility. Their response to encounter could be of vital importance. Consequently I have invited Duncan Lunan to detail in this chapter some chronicles of his content and discontent arising from the publication of his spaceprobe hypothesis. He has also contributed his opinions on the likelihood of SETI archaeology proving a rewarding course for investigation.

Ironically, those scientists who criticized his previous hypothesis did not challenge it on scientific grounds but on the basis of its philosophical naïvety. They should have been told that this man's training was not principally in science, as was theirs, but in philosophy.

Between 1967 and 1972 I chaired a series of discussions on interstellar travel and communication at ASTRA, the Association in Scotland to Research into Astronautics.* Our ideas were eventually published under the title *Man and the Stars*[1] (*Interstellar Contact* in the USA) and included a full statement of my own 'Epsilon Boötis' interpretation of radio echo effects in the 1920s, which, Professor Bracewell at Stanford University had suggested, might

---

*At that time the initials stood for Association in Scotland for Technology and Research in Astronautics. The name was changed for legal reasons at the end of 1976.

have been an attempt at communication by a spaceprobe from another civilization. The key parts of that hypothesis had already been published by the British Interplanetary Society and received worldwide publicity—much of it, unfortunately, sensational and inaccurate.

For the record, the outcome was that the radio echo effect was traced with virtual certainty to the orbit of the Moon; to one, if not both, of the 'Trojan' or 'Equilateral' points where James Strong of the BIS had suggested a 'Bracewell probe' might be found. However Epsilon Boötis, the star I had suggested as the origin of the hypothetical probe, proved to be twice as far away as most catalogues stated: it was too massive and therefore too short-lived to have sustained intelligent life on any planets. In any case, when more accurate records from the 1920s were unearthed, several of my crucial star map interpretations were proved invalid. As to whether a 'Bracewell probe' did or does exist in the orbit of the Moon—no experimental search was made and to my mind the question remains unanswered. A hypothetical natural explanation for the echo effect was advanced by A. T. Lawton and S. J. Newton[2], but I find it unconvincing.

The reaction of the world to those developments was both interesting and instructive. On the popular level, to put it bluntly, nobody wanted to know that Epsilon Boötis had been ruled out. Although the facts about the star and against my interpretation were known by the beginning of 1974, this chapter will be only the second time that the news has appeared in print. Like C. J. Jung and I. S. Shklovskii before me, on UFOs and the moons of Mars respectively, I find that the 'Lunan/Epsilon Boötis story' now leads a life of its own and is cited repeatedly by speculative writers to support ideas of their own. At least three such authors are currently in print with such claims *despite* having checked with me and having been told Epsilon Boötis was out. Many of the successful UFO hoaxes in the literature persist likewise despite efforts by their perpetrators to 'come clean'. Indeed, the discovery that the public and the media *want* contact situations to be true comes as no surprise: Jung[3] and Christopher Evans, in his excellent *Cults of Unreason*,[4] have explored the manifestations of that need, and more evidence of it will be given below.

The reaction from the scientific world was more revealing. First publication of the hypothesis had aroused little public comment, and what little hostile reaction there was privately was in response to inaccurate press reports rather than to my own paper. Most of the criticism concentrated on the impropriety of attempting to translate the 1920s signals, without prior proof that an extra-terrestrial probe existed; yet nobody was prepared to conduct an active search. The precise wording of three such refusals will be given below. Reaction to the publication of the Lawton/Newton hypothesis can be described only as unalloyed relief. 'Lunan's Theory "Junked",' declared a headline in the US journal *Astronomy*, without troubling to see if I accepted the refutation of the probe's existence. In fact I do not; but the authors of a recent book on CETI roundly declare 'Lunan now openly accepts Lawton's findings', an outright fabrication which they made no attempt to verify. Indeed, only two authors on this level have ever asked me whether I accept my supposed 'junking', and neither has quoted me as yet. The acceptance/rejection patterns, in the media and the world of science, are almost perfect mirror images; if the former reflects a desire for contact hypotheses to be true, there may be grounds for suspecting the objectivity of the latter. Stronger arguments are certainly needed to support that statement, and will be advanced below.

But the reaction from the general public, worldwide, was something else again; just as Carl Sagan has described the Pioneer 10 plaque as a 'a message to Earth' as much as to any future extra-terrestrial culture.[5] The spaceprobe hypothesis received press and/or radio/TV coverage on every continent on Earth except India (to my knowledge) and in most of the individual nations of Europe, the Americas, Africa and the USSR. To date the publicity and the continuing sale of *Man and the Stars* have brought in about 2,000 letters from all over the world, from people of all walks of life and standards of intellectual attainment, mostly asking for information. Now comes the rub: not one of those letter-writers was hostile to the idea of contact with other intelligence and not a single one was afraid of it. Many of the letters were foolish, some comically so, some sadly; but no one's folly or sanity made them ask me to leave well alone. Even those who doubted the spaceprobe's existence were at pains to say that they favoured CETI research in general, but there seemed to be far more who wanted reassurance (I use the word advisedly) that it could be true.

That particular outcome of the spaceprobe affair seems to me so important, as an impromptu opinion poll, that it deserves emphasis. It is, for example, a commonplace of the UFO movement that the world's governments suppress contact information 'in order to prevent a panic'. Well, nobody tried to suppress me, and there was no panic—far from it. In more rational circles, the scare caused by the Orson Welles *War of the Worlds* broadcast is often cited as proof of what would happen. Yet, as pointed out in *Man and the Stars*, that broadcast came across as live coverage of a particularly murderous invasion. In a lesser scare in London in the 1950s, a flying saucer had been described holding an atom bomb over the city—hardly an effective test of a true contact situation. At an early stage in the ASTRA discussions, the point was made to me directly: 'You think you'd be rational in the face of the unknown, but in a true contact situation you'd show the same unreasoning panic as anyone else.' Well, it seemed at the time that I had met the unknown, in the form of a star map drawn from the record of a radio signal, and my reaction was not fear but elation. If this was true, Man was not alone, interstellar travel was possible, logic was universal, communication was possible . . . 'But,' said my accuser, 'you're an intellectual, you're well informed on the issue, you're emotionally prepared for this to happen, your reaction won't be typical.' In fact, as the letters poured in day after day, it became clear that from schoolkids to convicts the reaction was the same: if the probe existed it was good news, and people hoped it was true.

In the scientific world, and among those who claimed to share its viewpoint, the reaction to the proposed search for the probe was either neutral or just the opposite of the public feeling. Not long before, Professor Zdenek Kopal of Manchester had published his memorable request: '. . . should we ever hear the "space-phone" ringing in the form of observational evidence which may admit of no other explanation, for God's sake let us not answer; but rather make ourselves as inconspicuous as we can to avoid attracting attention . . .'[6] One supporter of Professor Kopal was good enough to tell me that he would do all in his power to stop any attempt to contact the probe. Enquiries revealed that his powers were trivial.

Some scientists gave reasons for not answering the spaceprobe— pressure of other research, etc.; more often, correspondence was simply broken off without explanation; but in three cases of outright refusal, the precise words are well worth quoting. All three statements were made by gentlemen of considerable professional standing, but all were off the record, which is why they are anonymous here. The other factors they have in common should be all too readily apparent.

1). 'When we come across a phenomenon for which there is no natural explanation in terms of previously understood principles, we must not allow ourselves to consider the possibility that it may be artificial in origin.'

2). 'You must agree that it is better to do nothing about these things than to investigate them.'

3). 'Under no circumstances will intelligible signals be sent to the Moon Equilaterals, because that would constitute a biassed experiment.'

Such vehemence might be at least partly due to prejudice against space travel in general, and interstellar travel in particular. One of the gentlemen quoted above is a well-known opponent of manned space research, often quoted by inquisitorial senators trying to cut the NASA budget. Some time ago ASTRA was visited by an aspiring astronomer who said, 'You must agree that the Soviet results from Luna and Lunokhod are far superior to anything done by Apollo'—a statement palpably absurd in terms of quantity of samples, intelligibility of selection, recording of site before and after, distances traversed, range of terrain covered ... yet he was amazed at our reaction, since he had absorbed without question the idea that anything a man does a machine must do better. Fortunately NASA does not share that attitude and goes for results on all levels —practical, technical and scientific—rather than for 'purity' of method, despite the protests thus aroused. When Neil Armstrong said 'To hell with the international scientific community' it may have been tactless, but one suspects that he had been sorely tried.

In the interstellar field, there is still widespread acceptance of Purcell's 'all this stuff about travelling around the Universe ... belongs back where it came from, on the cereal box'[7] from the early '6os. The continuing debate among engineers on the feasibility of interstellar missions has made little impression: in 1973, reporting on the Cyclops study, *Nature* observed: 'Not surprisingly, it turns out that electromagnetic radiation is the best tool for the job.'[8] By that time, the BIS Project Daedalus team was well embarked on its study of an unmanned probe mission to Barnard's Star. Reviewing the second volume of *Interstellar Communication* (ed. A. G. W. Cameron) for the BIS Journal in 1975, Eugene F. Mallowe wrote: 'They seem to be unaware that a generation of engineers has been at work churning out innumerable papers purporting to show the feasibility of interstellar flight ... The category of interstellar transport and propulsion methods was regrettably omitted from the Bibliography by editorial fiat.'[9]

A deeper question underlies that debate—one which we had already confronted in the ASTRA interstellar project. One distinguished scientist (again I spare blushes) advocated firing cargoes of fertilized human embryos into space at random, most to be destroyed but one or two perhaps to be found and nurtured by civilizations elsewhere. This, he declared, would be 'philosophically more satisfying' than a purposeful programme of exploration and colonization. His hearers found no trouble in joining the Catholics present to say what they thought of his philosophy; but, moral issues aside, it seems perverse to the ideals of science to prefer a programme whose results are unknowable. Another scientist, no less eminent, told us that it was more sophisticated to communicate with other intelligences by radio than to launch manned interstellar missions. He quoted Purcell's analogy[7] of the picture gallery: it is more mature to look at the pictures and learn from them than to want to touch them. Again the society rose in protest—not just against the general principle, but also against the implication, false to scientific enquiry as we understand it, that it is more sophisticated to remain ignorant of the conditions in uninhabited systems. And, as Gerry Webb was to point out, reviewing *Interstellar Communication* for *Spaceflight*,[10] 'if the expected contact does not happen ... then after hundreds of years of first listening and then signalling fruitlessly, our (presumably by then) highly technologically competent civilization will be left with no alternative than to go and find out why no one answered our call'.

One of the oddest paradoxes of the 'Two Cultures' is that the philosophy of science is sufficiently difficult to be a final year honours *arts* subject, and therefore well outside the time commitment available to the science student. As a philosophy student with scientist friends I was aware of the problem, but the spaceprobe affair fully brought it home to me. One of its many aspects is the insistence on treating possible CETI phenomena as if they were purely 'natural' ('under no circumstances will intelligible signals be sent ...'), refusing to make allowance for the assumptions which might be made by rational minds at the other end ('. . . because that would constitute a biassed experiment'). It also involves the fundamental confusion between 'verifying' hypothetical processes involved in a natural cause-and-effect (e.g., 'the motions of double stars show that they conform to Newton's Law of Gravitation') and verifying hypothetical *facts* (e.g., 'there is a spaceprobe in the orbit of the Moon'). A sure sign of such confusion is a preoccupation with the 'respectability' or 'propriety' of the enquiry rather than with getting results.

It should be stated, then, that there is nothing in Western philosophy to which one can appeal to justify statements like 'It is philosophically more satisfying to send machines rather than men' or '. . . to communicate by radio rather than send ships'. As every first-year student learns, Western philosophy is descriptive, not prescriptive: it analyses the underlying principles of rational thought and ethical behaviour but does not provide precepts or a scale of values. In the more general everyday sense of the word 'philosophy' there is no value system I know which would make it better to look at a planetary system than to visit it; on the contrary, 'armchair traveller', 'bar-room mountaineer' and 'piloting a desk' are all terms of derision. Scientists stake a great deal on the purity of their ideals; where they clash sharply with public opinion, as in Professor Shockley's hypotheses on racial characteristics, the cry is 'we must go forward in the pursuit of truth without fear or favour' —a noble sentiment, however misdirected at times, and not one which leads logically to such propositions as 'You must agree that it is better to do nothing about these things than to investigate them' . . . nor to a scale of values on which unmanned probes or radio conversations are better than going to see for oneself. On space research in general and CETI in particular, one begins to suspect that something in the thinking is badly wrong. There are widely held views in astronomical circles that it is better in principle to explore the Solar System with machines than with men; that insterstellar travel is impossible, or, if it is possible, is undesirable, and either way should not be discussed; that it is preferable to converse with other intelligences by radio, with many years of timelag, rather than to meet them face-to-face; and there are strongly individual views that even such meeting of minds is to be avoided. If we strip those views of their supposedly 'philosophical' justifications—which are worthless if not harmful in this context—then we are faced with nothing more 'sophisticated' than an unrecognized fear of the unknown.

Carl Sagan has spoken at length about the various forms of 'chauvinism'—oxygen-chauvinism, liquid-water-chauvinism, ultra-violet-chauvinism, etc.—which in his view blind scientists to many possibilities for extraterrestrial life. Of all such blinkers, 'unknown-

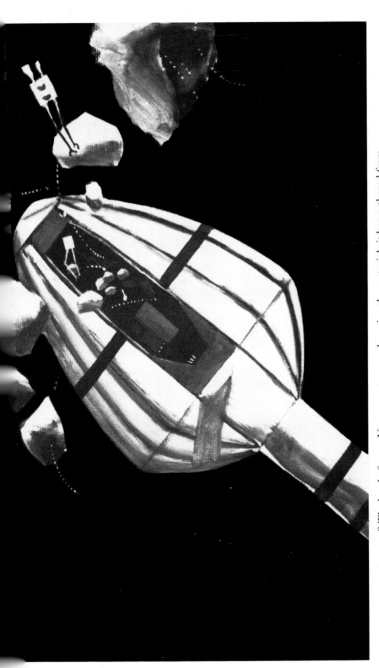

8 The head of a von Neumann-type probe using the materials it has gathered from the smaller bodies of the star system it is leaving to manufacture a new von Neumann-type (see page 114).

The detached head of a von Neumann-type probe left in orbit around an Earth-like planet. In the

chauvinism' would be the most misleading, time-consuming and wasteful. Wherever a line of argument was unconsciously recognized as likely to produce significant results the consensus of scientific thought would shy away from it, throw its energies into blind alleys, and denounce anyone who tried to open up the frontier marked 'Here be dragons'.

There is, of course, a large body of popular literature dealing with alleged past contact, virtually all of it sensational, inaccurate and misleading. Most of the remainder—books like Ronald Story's *The Space-Gods Revealed*[11]—have been written specifically to expose the phonies and are not themselves serious investigations of the past-contact possibility. But the odd thing about the apparent consensus in the scientific world is the condemnation not of the bad scholarship but of the *subject*. It is, for example, common practice to refer to the subject as 'von Dänikenism', although von Däniken was neither the first nor the worst writer in the field; but the label provides a convenient guilt-by-association implication that anyone who discusses any aspect of the subject must be equally sensationalist and inaccurate.

Reaction to the said crop of books, and particularly von Däniken's, is often given as a reason for refusing to allocate serious discussion to past contact. It should hardly be necessary, but unfortunately it *is* necessary, to point out that the reasoning is entirely spurious. In any other field of enquiry scientists would be outraged by the suggestion that a single sensationalist writer could muffle the entire scientific community and prevent all rational discussion. For example, I complained to the author of an otherwise excellent book on interstellar travel and communication that he had not mentioned past contact, even in refutation. His reply was: 'von Däniken has given the topic such a bad name that I could not risk giving it any space, for fear of cutting myself off from contacts whom I need in order to make a living.' If such voluntary censorship were the general reaction to outside speculation, science would stop dead. *The Secret Life of Plants* would make further progress in botany impossible; *The Bermuda Triangle* would bring oceanography to a halt; pyramid power would bring physics to a standstill and Velikovsky's absurd notions would stall astrodynamics for a generation. But I could not even convince the author in question that his scientist friends would not be *right* to take reprisals had he brought up this particular topic.

Mention of Velikovsky brings up the point that, in other fields, not only does science march on unimpeded by outside speculation, but it consciously shares the spotlight in the name of fair play. *New Scientist*, for example, remains proud that it gave Velikovsky a hearing, and cited that circumstance as a justification for a prolonged debate on Yuri Geller and paranormal claims in general. Not a year goes by without someone reopening the Velikovsky debate to make sure there has been no miscarriage of scientific justice—a weekend conference, sponsored by Glasgow University Department of Extramural and Adult Education, is closing as I write. And, while there is criticism of the methods and results of Professor John Taylor (on spoon-bending) and Professor Alan Hynek (on UFOs), nobody seems to question the rightness of enquiring. If 'unknown-chauvinism' is at work, of course, UFO investigations will be acceptable precisely because they are apparently a dead end, unlikely to reveal an encounter situation—present or past.

On the historical side of the topic, the pattern is likewise all too familiar. I recently had occasion to ask a well known researcher if he had come across any oddities in a certain field, perhaps indicating the presence of visitors? No, he replied, 'but in any case, that whole class of enquiry is thoroughly bad scholarship'. He was objecting not to a particular hypothesis or approach, but to the entire topic, in which no form of enquiry could be valid. ('If we come across a phenomenon we cannot explain . . . we must not allow ourselves to consider the possibility that it may be artificial . . .')

Likewise, I was drawn some time ago into a correspondence in the SFWA *Bulletin*. Replying to my letter, L. Sprague de Camp wrote, 'To see the myth of the extraterrestrial enlighteners de-bunked once and for all, I refer Mr Lunan to *The Space-Gods Revealed*, by Ronald Story.' Story himself makes no such sweeping claim: his target is announced in his subtitle: 'A Close Look at the Theories of Erich von Däniken.' He states in the Appendix that he has not made a detailed study of (equally bad) books by other writers, which were consulted only to gain a general impression of the field. And even of von Däniken's best known claim, that astronauts created mankind (for which he offers no hard evidence, and the hearsay evidence is all spurious), Story writes: 'This hypothesis, while quite unlikely, is not logically or physically impossible; in fact, one underlying supposition—that intelligent life may exist elsewhere in the Universe—is now denied by practic-ally no one who is reasonably well informed on the subject.' Story goes on to a detailed breakdown of von Däniken's faulty evidence and reasoning but undertakes no discussion of the possibility that Earth may at some time have been visited. Far from ruling out the possibility, Story does not in fact discuss it directly at any point in the text.

In this example and the previous one, two distinctive symptoms of 'unknown-chauvinism' might be detected: concern with the supposed impropriety of the enquiry, rather than with the facts, in the first case; and accepting that any attack on any part of the subject closes the whole enquiry 'once and for all', in the second.

The emphasis on propriety rather than truth is particularly worthy of attack. It contradicts the scientific ideal that any question, no matter how startling, may be asked without fear or favour. But it also highlights the logical distinction I drew between testing a hypothesis, in the scientific sense, and establishing a fact. For example, Newton's 'Law' of gravitation was a hypothesis which was considered to be verified because it correctly predicted observable phenomena, within the limits of measurement of the day. In time more subtle observations revealed discrepancies for which Einstein's more complicated theories were able to account, in addition to satisfying new tests of prediction. Newton's account of gravitation has not been invalidated—it remains 'true' as a description of certain specialized cases within the more general Einsteinian framework—and we cannot rule out the possibility of further evolution of our insight in the future. By contrast, either the Earth has been visited or it has not, and it should be possible in theory to answer the question with a definite 'yes' if true. (A definite 'no' would be harder to obtain, but if it is the case we should try to establish it as firmly as possible. The most likely implication, that we are one of the first high-technology civilizations to migrate into space, is not exactly trivial.) The factual knowledge gained from archaeological discovery is not provisional, in the sense that any theory of physics must be: after Schliemann found the ruins, the existence of Homer's Troy could no longer be queried in the way that theoretical challenges are periodically aimed at Einstein. To say that all past-contact enquiry is bad scholarship is like saying Schliemann was wrong to look for Troy on the basis of myths— even though Troy actually existed. In chauvinist terms, Schliemann's dig was unquestionably a 'biassed experiment'.

In *Intelligent Life in the Universe*[12] Carl Sagan suggested three criteria for identifying a possible contact situation in the past. These were: (1), an account committed to writing at or near the time; (2), a major effect on the civilization contacted; and (3), no attempt to conceal the extraterrestrial nature of the visitors. I have discussed this problem at length with a former intelligence officer, Alan Evans, and in our opinion Sagan's criteria are too demanding in some respects and too lenient in others: e.g., the contacted society might not have possessed writing at the time, and might have insisted on deifying the visitors; they might not have recognized the extraterrestrial connection, especially if they did not witness the landing; nor is it hard to think of reasons why visitors (well meaning or otherwise) might decide to accept deification in the circumstances. There are in fact no alleged instances of contact which wholly satisfy Sagan's criteria; but it seems to us that on analysis they serve primarily as guidelines to preserve the reputation of the investigator, in relation to specific alleged instances, rather than to answer the question 'Has Earth ever been visited?'

In the course of the spaceprobe affair I evolved four criteria which I believe to be more helpful. When considering past evidence as possible indications of Contact, we should look for:

(1) recognizable principles of technology;
(2) rational purposes for the supposed Contact;
(3) information freely given;
and, resulting from these three,
(4) further possible investigation capable in theory of verifying or disproving the supposed Contact.

It should be noted that the postwar hordes of UFO sightings, including alleged meetings with UFO occupants, have failed to meet these criteria singly or in groups. From the viewpoint of an investigator in a thousand years or more, however, it might seem that they satisfy Professor Sagan's.

Criterion (3) is the one which may not be satisfied. It is easy to imagine that visitors might choose not to give information, especially if they were conducting an investigation prompted by conservationist motives. If that were the case, however, presumably they would also be systematic about removing artefacts and our chances of verifying their presence here would be remote. What follows is based on the assumption that we are intended to find out, sooner or later, who came here (if anyone) and why.

Almost certainly, extraterrestrial artefacts are the only acceptable category of proof that contact has in fact taken place. It is hard to imagine any other category of evidence—no matter how good the instances within it—which would carry sufficient force to make its acceptance anything but a matter of opinion. Artefacts can be viewed in four divisions: objects deliberately planted, objects abandoned, objects on the Earth, objects elsewhere in the Solar System. Objects elsewhere in the Solar System are not likely to be found, at the present stage of exploration, unless they deliberately attract attention by responding to radio signals received from Earth. One of the attractions of the spaceprobe scenario was that it seemed to be a positive instance of this kind; it is also one of the few which could theoretically be verified without extending our present space programme.

Objects on the Earth could fall into an almost infinite range, but for our purposes can be classified as: (a) objects deliberately left to be found by us (information, relics, monitoring devices or communications systems); (b) objects deliberately concealed from us (monitoring or other devices serving extraterrestrial purposes); (c) objects discarded; (d) vehicles or installations wrecked or abandoned.

Category (a) objects are meant to be found and presumably signposts exist. Category (b) ones may give themselves away, if still active, if we now have the technology to detect them and their makers have had no opportunity to switch them off or remove them. Category (c) seems unpromising until we know what happened and where it happened. At first sight category (d) seemed most promising, but the hope was not borne out. Unless a space vehicle got into difficulties near the ground, or was overwhelmed by some natural catastrophe like a storm, earthquake or volcanic eruption, there is little chance of leaving recognizable wreckage. At an early stage in our correspondence Alan Evans suggested surveying catastrophe sites such as Santorini in search of wreckage, but the chances of finding anything are not good. According to a Rolls-Royce technician we consulted there is not much chance of recognizing high-technology equipment, after thousands of years in the hostile free-oxygen, liquid-water, high-gravity environment of the Earth's surface, unless deliberate steps have been taken to preserve it.

Thus the probabilities come down with some weight in favour of category (a), assuming that the assumption of criterion (3) is correct. Our best prospect of success is to search for artefacts left and signposted for us. On that basis there could be four classes of possible evidence.

(A) Artefacts of incontestably extraterrestrial origin—data capsules, spacecraft, monitoring devices or installations.

(B) Photoreconnaissance or electromagnetic anomalies leading to the discovery of class (A) evidence.

(C) Clues built into long-lived man-made structures (particularly astronomical clues) leading to the location of class (A) or (B) evidence.

(D) Legendary, mythological or archaeological material (man-made) reflecting some aspect of a past contact—if one occurred. All the 'von Däniken material' allegedly falls in this class, but in fact very little of it stands up to preliminary investigation and the remainder is far too tenuous even to be classed as 'evidence'. The only possible uses for such material are to indicate possible historical and geographical areas for investigation in the other three classes, or for possible correlation with that evidence if found.

In *Man and the Stars* I attempted a preliminary analysis of the class (D) material advanced by von Däniken in his first two books, plus seven books by other authors. The result carried so little weight that it had to be presented in humorous terms and inevitably was coloured by the Epsilon Boötis scenario, now discarded. Even so the most favourable conclusion was that the Earth had not been visited more than four times. Further research has not extended those possibilities: the general trend has been to strengthen the alternative possibility that Earth has not been visited more than once (if it happened at all), with 'frozen' memories appearing in the folklore of later cultures. There does appear to be a possible correlation with class (C) evidence, if valid; but of course coincidence cannot be ruled out.

Thus we come down at last to class (B) as the make or break point. The only effective way at present to find artefacts, prove Earth has been visited and establish that we are not one of the first civilizations in our part of the Galaxy, is to look for photoreconnaissance or electromagnetic anomalies. The precise nature of the traces, if they exist, will depend on the technology and purposes of the visitors, and rather than spell out a list of possibilities, it would be better to rely on the photo-interpreter's judgment or intuition that he has found something unusual. But, since Earth is under intensive study from the air and from orbit for military and civilian purposes, it seems that the cost of such an investigation would be limited to the gathering, interpretation and investigation of possible instances. Some thought has been given at ASTRA to the handling of such a project and we will be very interested to hear from anyone in a position to set it up—or even to help in pursuing class (c) and (d) leads with photographic evidence. But it must be emphasized that the object is to locate class (a) evidence if it exists—not to launch a new set of sensational red herrings in class (d). On the other hand, such a search seems the best chance before us at present, if we are not alone in the Universe, to establish the facts in our lifetimes.

8

Meditations of the Heart

Artistotle taught that the seat of the emotions was not the brain but the heart. He was wrong. Anyone who made a statement to the effect that the brain was the seat would be nearer the truth but they would have a too selective view. The true seat is a particular interpretation of the entire human body which takes into account not only the nervous system but the hormonal balance and much more.

By emotional behaviour most people mean irrational behaviour. This begs the question. What is rational behaviour? Is it activity directly and only related to reasoning? Is it, on the other hand, related to the sociological imperative created by us and dictating to us? How do reason and emotion relate to empathy, imagination and curiosity?

Do they relate at all to the ETI?

Take curiosity, an aspect of terrestrial behaviour, a seeking out of new significances within the Universe as we interpret it. We do not know that curiosity is the only basis for activities which make beings elsewhere aware of new significances. If curiosity is truly the psychological foundation for scientific activity then we cannot expect the scientific method to be widespread among ETI. This is not to say that high mental lifeforms will prove to be incapable of activities which we might interpret as being scientific or technological in their nature. The problem lies with our very interpretation of these activities.

We must be prepared to broaden the nature of what we consider to be science, technology and perhaps even curiosity itself if we are to relate them to ETI behaviour. For example, we are unaware of how it would be possible to cross interstellar space without a sophisticated grasp of mathematics and logic which we employ in system designing. This does not bar the almost certainly non-mathematical AMLs from moving between the stars. They too will have to employ sophisticated use of their 'brain functions' as will we with ours in order to do so. We cannot do it without maths: this is a description of our limitations, not of theirs.

Let us recap on the possibilities seen by most CETI scientists searching for radio signals. Firstly, there may be a beacon which is attempting to set up a communication link with as yet undetected civilizations. Secondly, there may be the possibility of us intercepting a message being transmitted from one ETI to another. Finally, there is the chance that we may pick up the expanding sphere of internal communications of some fairly nearby system—just as they may pick up the internal signals which have been expanding from Earth for decades now.

The first two seem somewhat anthropocentric in their conception and the latter may prove totally incomprehensible. It may well be the radio-communication 'songs' of a mental lifeform which uses radio to communicate in the same way that our cetaceans use sound. If this is the case, then we will be on a sounder base for comprehension once we discover and begin understanding the world of dolphin significances.

There is a possible other form of signal which may be only marginally associated with communication. Something of the nature of an interstellar radar pulse may 'hit' the Solar System to read it for aspects which the ETI would find significant. The pulse may not be comprized of one modulated signal but of many of many different kinds on a variety of wavelengths only some of which, if any, will correspond to radio frequencies. Should we intercept the signal it will give us information, no matter what type of signal it is.

Let me dwell for a moment on the idea of the sophisticated 'radar' pulse. Radar is an inappropriate description for the exploratory device suggested by Dr Alan Cottey[1] of the University of East Anglia. Somewhere in between the use of radar and a miniaturized spacecraft for the purposes of interstellar exploration there may exist highly versatile non-linear but wave-like phenomena capable of travelling at near light speed and of interacting in a highly sophisticated manner with their target environment.

Another suggestion is for the transmission at near light speeds of biological materials which then interact with the target system to produce a variety of biological machines capable of exploration and returning data to the home planet without actually returning physically themselves.[2]

Perhaps the two suggestions could even be married in a number of ways. The pulses could themselves be so highly organized that they constituted a high mental lifeform in themselves; a point which Alan Cottey has underlined with a warning that it could be folly to attempt to make distinctions between the ETI and its artefacts. Alternatively, the pulses themselves could produce the biological machines.

Whether or not they would actually be mental lifeforms is a moot point with regard to such sophisticated signals. If they were so highly organized that they did not simply bounce back an echo but the whole signal-pulse package system was capable of returning highly processed data to its source it would be of the stature at least of a superb terrestrial computer if not actually of an artificial intelligence.

Some brave individuals 'guesstimate' that there are 'civilizations' emerging in the Galaxy at the rate of one every decade.[3] Let us substitute ETI for a 'civilization' and we may be on the way to establishing a better basis for relationship between ourselves and other mentation in the cosmos. Being completely unable to appreciate the motivations of the ETI we will be totally at a loss to understand the prime movers of their interstellar urges. This being the case, we cannot be sure that they will be interested in the dissemination of knowledge.

Given that there is no way of determining the motivation behind the ETI's interstellar urge, we cannot anticipate what strategies will be considered and employed to satisfy it. This being so, we should regard all strategies as being equally likely. The more strategies which we dream up to satisfy our own urge the less likely we can consider any individual one as being the best for the ETI.

What we must look for is the strategy which gives the best prospects for encounter under these conditions. There are two imperative bases for such a strategy. The first and most important is that it must employ a system which is not solely rooted in our own mathematics and technology. The second is that, even if it is adopted only very rarely throughout the Galaxy, say once every one or two million years, and the originating AML soon loses interest in it, we still have a fairly good chance of encounter.

Both of these conditions exclude radio communication: the first because, although there are natural radio transmitters galore and it is simple enough to imagine an ETI with a natural radio ear, the codes would have to be mathematical before we could understand them (although computer analysis may show statistical improbabilities in some and indicate that they may be ET in origin); the second excludes it because, if an ETI loses interest in radio signalling rapidly, we simply do not have time on our side.

The one concept which holds out promise to me is that of the renewable space probe.[4] This is basically a system which is capable of reproduction. The mathematician John von Neumann took the concept of a machine which could duplicate itself and worked it into a precise mathematical description.[5] A system of interstellar exploration based on such machines would be very efficient indeed.

Von Neumann's work anticipated to a remarkable degree what was subsequently discovered about the ways in which known living systems reproduce and repair themselves. This is what I find fascinating. The ant with its lack of mathematical or biological science reproduces heartily, as do whales and viruses. An ETI would have to employ such a system within its own frame of significances but the system itself would not be rooted in these any more than it is in our maths and technology. We cannot say how they might employ this quasibiological system, but we can make guesses about ourselves.

Assume that in a century hence we have the capabilities to build both a von Neumann-type machine system and an interstellar probe. There is then the possibility of constructing the self-reproducing probe-system. What must be recognized is that if this endeavour is embarked upon it will be enormously expensive, much more so than a non-reproducing probe. The great advantage of it is that once the vehicle has been built and is fully operational all our capital costs on probe building are over. For that investment we have a system which can explore every star system, every interesting astronomical object in the Galaxy.

Who can say what our technological capabilities will be a century hence? It may take much longer or much shorter, but there are a number of elements which have to be incorporated into the design. The probe will have to be a system capable of replicating itself several times over in almost any environment it is likely to encounter, perhaps even interstellar space itself. It must be able to communicate with Earth and other probes. One of the problems which we can currently foresee is that of a propulsion system. Hopefully, as we already have a fair design for an interstellar probe at the present time, a variety of propulsion possibilities will be both evident and in some cases actually in use by the time of the von Neumann probe's construction.[6]

By its very nature the probe will be able to process all its own requirements over a very long period of time. This means the machine will take care of providing itself with suitable materials and fuel for the interstellar journey and its self-maintenance at the other end.

The process would work something like this.

A fully automatic probe factory would be set up, possibly in the asteroid belt or in orbit about Jupiter. This would assemble probes and despatch them. The exploration strategy would probably be fairly straightforward. Probes would be sent out to the nearer stars continually until the returning signals indicated that they had all been successfully 'colonized'.

Probes *en route* to stars successfully reached and monitored would then select their secondary targets. These would lie beyond the immediate group that the Solar System factory was interested in. They would inform the factory and other probes in their vicinity of the choice. This expanding net would be passing back to Earth data concerning the interstellar medium through which it was passing.

On arrival in a target system the probe would seek out some suitable site and lay its 'egg'. This would be a complex of computer elements and handling mechanisms. This system would then set about creating more elements which it would add to itself in a cycle of enhancement until it had fully developed into another automatic probe factory.[7] As soon as the first planetary probes had been successfully deployed the factory would signal Earth that the star system was being monitored. The next task would be the building of a batch of interstellar probes each similar to the one which was the factory's parent.

The planetary probes would be self-repairing and self-maintaining to a great extent but, when the maintenance and repair systems finally broke down, the factory would produce others. As the factory approached the end of its own lifetime it would create a quasi-star probe not designed to travel from this system but merely to lay a new 'egg' from which a new factory would spring.

It might be relevant to give the planetary probes a capability to produce such a quasi-probe among themselves. This would be the equivalent of a beehive selecting a female to replace the queen when she dies. Just as special procedures can turn this female worker into a queen, so might the probes turn one of their own into a quasi-probe for the purposes of 'egg' laying. This would ensure long-term monitoring of not merely the system but whole sections of the Galaxy and eventually the Galaxy as a whole.

With high-quality inter-probe communication this system should, after a great length of time, have explored the whole Galaxy and be in a position to send new probes to emerging stars as well as to the Magellanic Clouds and other extragalactic clusters of stars.

This is the von Neumann spaceprobe strategy as viewed from the human point of view. We know that we would like to explore the Galaxy. We may well decide to do it in this manner or we may not. The important point is that, as far as the ETI is concerned, the von Neumann probe is a more likely strategy for encounter.

My feelings are that it would be very surprising if we had not been visited by a system in some respects similar to this. Indeed, if the profusion of 'civilizations' which have possibly come into existence since the birth of the Galaxy some fifteen billion years ago is anything like that which the CETI investigators believe, then there may well be several systems in our vicinity—and active ones at that!

Beings, or a being, responsible for satisfying their interstellar urges in this way will not necessarily be interested in CETI. In fact, the von Neumann-type spaceprobe would be of value in CETI terms only to a civilization which was much more patient than our own appears to be. The length of time which would be needed to set up transmitters in the star systems within, say, one hundred light years of the point of the probes' origin is much longer than that needed to investigate the same number by radio (either 'radar' or simply conventional listening techniques).

If von Neumann-type probes have been entering our system the 'lifetime of civilizations' condition which should interest us is different from that used in conventional CETI equations. (The most interesting type of probe is going to be an active one. Only a functional probe can respond to a systematic search, and it will be much easier for us to detect a responsive artefact than a dead or totally uninterested/passive one.) The factors which affect this are, then, the average number of years between civilizations adopting this strategy, the proportion of such explorations which make a successful sweep of the entire Galaxy (and thus enter the Solar System), and finally the average length of time probes or their descendants remain active in the target systems.

An interesting point here—particularly with a view to our educating ourselves in the finer points, or even the basics, of CETI proper—lies in the possibility mentioned above that probes from differing systems may coexist in one target area. Although we cannot assume that they would be even remotely interested in one another it is exciting to speculate what the results would be if they were.

Such a situation could bring about one of the great clichés of science fiction—the *Encyclopedia Galactica*.[8] Let us imagine that, over the billions of years Advanced Mental Life has been appearing and perhaps disappearing in our Galaxy, there have been a few encounters between von Neumann probes (vN) interested in an interchange of information. The data which they would then compile would illuminate the different outlooks and different psychologies behind the construction of the probes.

In this situation there would be some higher degree of correspondence. Both of the ETIs responsible for building these probes would have certain areas of their mental life aligned if their probes, which they designed and programmed, elected to follow this common course. The first and most obvious factor which they should have in common would be the use of this particular strategy to satisfy their interstellar motivation. Secondly, and for our purposes more importantly, they would be probes capable of being motivated to attempt CETI. As long as they or their descendants remained active, which may run into many millennia, they might be able to tackle the problems of communication between mental lifeforms at least on their own level. Nor need the communication attempts be worked on in only one star system. What we would have would be two areas of expansion overlapping to an increasing extent. In consequence, encounter between the probes will take place in more and more systems. If one species of probe has a system of information-exchange among its own members, the first encounter could be made known throughout the species as rapidly as relay permits. In this case each probe which decided to attempt CETI would begin searching its target system for evidence that a member of the other ETI species was present there. Even if none were present, there could still be theoretical work on CETI which the probe might do using data transmitted by other probes, several light years distant, too heavily loaded with immediate practical work to take it on themselves. The fact that such probes may be many light years distant from their peers is not a great barrier in this case. The probes working on CETI will have hundreds of years to tackle the immediate interaction with the other CETI-oriented probe species. Lifetimes may run into thousands of years for the various generations. They can afford a degree of patience which we humans cannot. Their clock is not the twirling Earth but the revolving Galaxy.

With the interaction between the two species taken as far as it can go there may then be an exchange of information in which the two, with their different significance bases for interpreting phenomena, will be able to enrich one another's picture of the Universe.

There is no way of projecting how long a community of von Neumann probes could exist in the Galaxy. If, for example, there was a local disaster in one target system which destroyed the 'factory' and 'monitors' there it would be a simple operation for a vN in a nearby system with which it had been exchanging information to dispatch aid in the form of an 'egg'-laying probe.

So far I have discussed the probe species as having lifetimes of some thousands of years, but in the above case lifetimes might well run into many millions, perhaps even tens of millions of years. If the probes were capable of some redesign, they might also improve their survival characteristics by producing carefully considered mutations.

Should an ETI adopt the vN strategy once every million years there is still a chance that two probe species might interact. The likelihood of two CETI-oriented probes interacting decreases with the average length of time between their appearance. However, even without effecting CETI, an exploratory probe species may have a large databank concerning the nature of life in the Universe after having observed numerous star systems over a period of some thousands of years, and this may well include studies of other AMLs.

If they exist long enough they may well have a highly developed ability to identify and track other ETI probes, not necessarily von Neumann in nature, as they approach their target system. In this case the original contact between the two CETI-oriented probes may well be initiated by the already on-station probe in the target system as it observes the new one entering.

One fact, although not a particularly significant one perhaps to the observing ETI probe, is that a great deal of energy is generally required if a probe is to be accelerated out of its home system to an interstellar escape velocity. Similarly, a great deal must be used to decelerate it into the target system. One strategy which has occurred to a number of mere humans is that we may try to detect the approach of probes entering our system. Although no ETI probe will have human-based programming it may use a strategy for probe detection. In this case an entering probe will encounter evidence for ETI in the target system should the vN already resident there try to contact it. These vNs will not be indigenous to the system but the entering probe will not be aware of this. If it is a comparatively primitive probe from a home system lacking a vN 'colony' this may precipitate some peculiar conclusions. The builders in the home system would receive quite a shock. In the Daedalus scenario it would be the equivalent of our having an ETI message relayed to us from a star system right on our doorstep, and the repercussions would be dramatic.

Various factors influence the way in which different species of von Neumann probes might interact with one another. The crucial ones, as I have mentioned above, are how often a species of probe appears in the Galaxy and how its reproductive programming governs the way it spreads. If species die out, or are deliberately closed down, after an average period of time which is less than the average period of time between their appearing, meetings would be very few indeed.

Another factor lies in the possibility that, if the probes are fairly flexible creations, capable not only of reproduction but of some redesign, they themselves may be subject to a long-term developmental process akin to evolution. Their original creators, the beings who had the initial interstellar urge, would continue themselves to be subject to the evolutionary (or evolution-equivalent) processes which brought them into existence in the first place.

It has been pointed out that ETIs interested in radio communication may cease to be interested either in the subject of CETI or in using radio as they climb higher in their development. If the probes which they send out mutate in such a way that they perform their initial tasks with greater efficiency, then the problem of probe mutation is one which works in favour of CETI, in a CETI-oriented probe, and heightens the chances of both survival and communication. Should the probe design mutate in some way which is not concerned with increased efficiency there is a chance that some species will lose interest in CETI. Hopefully, this is balanced by the emergence of a mutated interest in it among probes not intially programmed to cater for the concept of ETI at all.

Now let us look at another extreme. What if the average lifetime of a probe species is a great deal longer than the average time lapse between new species appearing in the Galaxy, and if most species are interested in CETI to some degree?

10 A von Neumann-type probe enters a new solar system (see page 114).

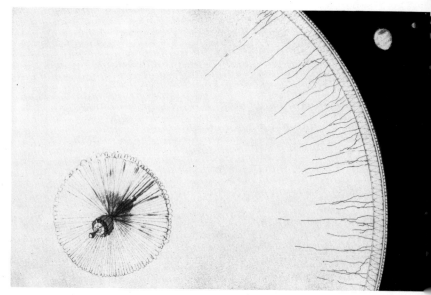

11 A subprobe detached from a von Neumann-type factory. It is using the solar sail principle as a means of interplanetary propulsion.

12 A 'college' of von Neumann probes exchanging information (see page 118).

Let me illustrate this with the possibility that it need take a species of von Neumann probes no more than three million years to saturate the Galaxy. If the average lifetime of a vN family about its target star is in the region of five million years then we have a rough total of some eight million years during which the species exists and a period of two million during which communication can take place between its most distant members. In this scenario there would of course be only ten such two-way communications between these distant relatives. However, the entire tribe would number around 120,000,000,000, or a minimum of one for every star in the Galaxy. They would be handling quite enough communication from the family members among their neighbouring stars to be little concerned with sending messages right across the Galaxy.

Let me now be madly optimistic and assume that one mental lifeform every 50,000 years adopts a von Neumann probe strategy with an interest in CETI. In this case, between the birth of one species and its demise 320 others will appear, of which the first 40 will share Galactic saturation with the first probe exploration.

There is no scientific basis for choosing the above figures;* they merely represent a very optimistic scenario on my part for CETI. The ultimate in optimism, of course, is to believe that the quoted figure of one 'civilization' emerging each ten years also applies to the eventual emergence of a von Neumann probe from that 'civilization' and that even on the cosmological scale the probes, because of their reproductive capability, are virtually immortal.

Should ETI have appeared prior to the formation of the Solar System it means that a vN may have observed the entire history of our Sun and its entourage of planets, and that there could be as many as a million-plus 'dead' species in our vicinity. Of these, around one hundred thousand would have contributed to a vast body of knowledge about our immediate stellar and interplanetary environment.

If ETI interaction has been taking place for billions of years right on our astronomical doorstep we do not have to start searching among the stars for evidence of their existence. It seems fairly sound for us to assume that they are quite well informed of *our* existence. If they wish to communicate, their expertise in handling CETI problems will ensure a fairly sophisticated sequence of opening moves on their part.

*See Appendix 3.

So much, alas, for optimism. The reality is that we have no certain knowledge of anything concerning advanced mental life elsewhere. No ETI may *ever* adopt vN as a strategy for Galactic exploration and, even if they do, they may not be even remotely interested in CETI. It must be said, however, that the same applies to radio broadcasting/listening techniques: essentially the ETI is an unknown quantity. However, the important aspect about the vN strategy is that the concept is very broad-based. From our point of view it would be a Galactic exploration programme with, one hopes, a CETI factor of some nature incorporated. That is, if we ever built one—and of course we have to build only one.

The basic problem with the CETI scenario I am setting up is 'first catch your vN . . .' To take an extreme case, suppose the average vN totals some twenty thousand tons, arrives every decade and reproduces completely once every five hundred years or so. In this situation they have eaten up less than .002% of the Solar System's planetary mass so far, assuming they have been popping in regularly during the entire existence of the System. To complicate matters, they probably do not arrive so frequently, reproduce at a more leisurely pace and 'cannibalize' the defunct elements which are being replaced. It follows that their mass is not going to be of particular assistance in determining where they are, and certainly not in seeing them.

Accidentally overhearing their domestic communications is another unlikely factor. If two monitoring elements are corresponding they will know the subjects of their monitoring well enough, namely the components of the Solar System, to make sure that none of them wanders in the way. A vN devised by ourselves would probably incorporate some kind of central switching technique, like a telephone exchange for spaceprobes, and this would handle all non-emergency data traffic.

For all we know, they may even have tried to contact us and we have simply ignored them. On the other hand, why should they bother? Their makers had a higher standard of comparative capabilities than we currently have. To them we must be among the less highly developed mental lifeforms. Perhaps, if this is so, they will attempt to communicate only, if at all, when we activate our own vN.

Despite all these drawbacks I still think that searching for vNs in the Solar System is more readily justifiable than radio searches for interstellar communications. Let me explain why.

Assume for the sake of argument that over the next few hundred years Man's exploration of space is purely by means of robot probes to observe various elements within the Solar System, and that his CETI endeavour is literally channelled into the realms of the interstellar signal. If we do not hear anything over the first ten years the hypothesis is still not disproved: the signal may come during the next ten years. If we have heard nothing at the end of the first century some scientists will be having grave doubts about their original assumptions concerning the rates of emergent 'civilizations' in the Galaxy which reach the stage of radio communication. These doubts will grow, and greater revisions will be required if nothing is heard after five hundred years, and more so after five thousand. *But the absence of a signal has not disproved the original hypothesis because the signal might always arrive tomorrow.* We could go on squatting here on planet Earth, listening vainly forever and still not disprove it. There is no way of falsifying this proposition. It is the perfect example of an *ad hoc* hypothesis.

My CETI hypothesis, the vN hypothesis, is also *ad hoc* at present but within the coming years it should prove possible to disprove it by scouring the Solar System to establish the absence of the vN. Dare I suggest that we might try out the Lagrangian points in the Moon's orbit about Earth using a lunar orbiter and, down here, a directional radio transmitter?

At this particular juncture I would like to throw my contributions into the speculations regarding the possibility of there being a probe in the Lagrangian positions. There seems to be a great deal of muddy thinking taking place in this area. I hope I can make things a shade clearer and not darker still.

If we wish to set up a postulate regarding a probe in these positions, we must work within consistent references. A hypothetical situation must be shown to be false within the terms which it proposes. This has not been the case regarding the Lagrangian spaceprobe hypothesis. A recent theory has been put forward giving a natural explanation for Long-Delayed Echoes:[9] this, however, is not relevant. It is important to understand *why* it is not relevant. The existence of a spaceprobe does not *ipso facto* rule out a natural explanation for LDEs as there is no reason why natural and artificial phenomena cannot exist side by side. As a consequence, the postulate of there being an extraterrestrial probe out there cannot be 'explained away' in these terms. Explaining LDEs in natural terms is not a fruitful course of action regarding the spaceprobe hypothesis: it is concerned with natural phenomena only and is therefore useful (and perhaps extremely useful) to those who seek to know more about the physical characteristics of radio in the near-Earth environment. Such characteristics cannot disprove the proposition that there is a probe at one of the Trojan positions.

To do so we must tackle the hypothesis within the terms which it proposes. The most obvious strategy—and the most immediately difficult—is to send spaceprobes into those positions to give us detailed knowledge of them. It would be ironic if, after flying the multi-billion dollar Vikings to Mars in search of microbes (among other things, of course), we discovered that by sending little more than a TV camera-plus-transmitter 0.1% of the distance Viking travelled we found evidence of not merely life but advanced mental life.

Hit the Lagrange positions with radio signals. If no response comes to a wide variety of signal types over a few weeks or at most months simply pack up. The probe is either not interested, not functional, or simply not there. In any event the proposition that a radio-contactable probe was hanging about at those points just fell over.

The overriding virtue of the experiment is that it is astoundingly cheap.

Should vNs be here, the factory elements are probably in the asteroid belt, somewhere in the vicinity of Jupiter or perhaps fairly close to the Sun. The latter could act as a kind of massive catcher's mitt for an incoming probe: the wind of electrically charged particles issuing from our star might be used for some form of magnetic braking. This would possibly result in an orbit close to the Sun, from which the factory 'egg' might mine its materials.

Perhaps they might take up residence on comets, doing a sweep of the system every few decades to check up on their interplanetary progeny. Then again, a suitably altered asteroid, engine fitted, might coast and cruise among the planets as a supply-cum-factory vessel.

The possibilities are quite intriguing.

9

Unbelievably Fatuous Observations

Within the terms of the vN hypothesis it is tempting to assume that certain probes might manufacture small monitor vehicles with atmospheric capabilities which could observe planets very closely. (UFOs!) On the other hand, the vN hypothesis is sufficiently *ad hoc* as it stands without encumbering it with any further paraphernalia. UFOs are certainly phenomena eminently worthy of investigation. The propositions that they are extraterrestrial vehicles, or terrestrial time-machines from the future, or disembodied intelligences from other worlds/dimensions, etc., etc., are great fun and earn some people a good deal of money, but how useful are they? There appears to be a great deal of imagination at work here and very little insight.

As a classification of most UFO reports my own listings are rather different from Dr Hynek's.[1]

First Kind: misinterpretation of natural phenomena such as planets, stars, the Moon, meteors, clouds, birds, freak reflections, 'mirage'-producing temperature inversion layers, freak electrical discharges in the atmosphere, fireball-plasmoids, etc.

Second Kind: misinterpretation of artificial phenomena such as aircraft, balloons, fireworks, flares, spotlights, satellites, even ground and sea transport viewed at night in unfamiliar terrain.

Third Kind: conscious deliberate trickery either on the part of someone who wishes to convince others that he or she has experienced a 'close encounter', or someone wishing to convince others that they are experiencing or have experienced one; honest mistaken belief by individuals that they have experienced something which they have not.

There can, of course, be a combination of many of the elements listed above to convince people that they are witnessing an extra-terrestrial visitation. The most convincing UFO reports may be those containing such a combination. The fact of UFOs being reported is indisputable as is the fact that by far the majority of these reports is satisfactorily explained. UFOs which remain totally unexplained are *ipso facto* unidentified. This means that it has been impossible to identify them as anything—including spaceships.

No one with an iota of wonder would doubt that our experience is full of indications that the Universe is a very weird place. What we must be wary of doing is placing preassessed interpretations on the various weird occurrences. The approach which I would like from a ufologist (*sic*) is a more critical one. Should a statement be made concerning ways in which I or anyone else could set about *disproving* the proposition that UFOs are spaceships I, at least, would be very interested.

There is always the possibility that some UFOs are of artificial non-human origins. It niggles away in the attic of my mind but there is nothing useful that it says to me about CETI; in consequence, I let it niggle on up there among all the other *ad hoc* hypotheses which I only let out on the strict understanding that they are going to behave themselves. As soon as bright coloured lights start sweeping across Arizona playing music scored by John Williams I am going to look very foolish indeed.

So is there anything useful which we can abstract from the contents of the whole Flying Saucer bandwaggon? Certainly there is nothing that is going to throw up insights into the nature of the ETI. On the other hand, what about the nature of mankind?

There is a wide and commercially well exploited willingness to believe in ETI on the part of the public. If tomorrow I decide to write a supposedly autobiographical book entitled *My Experiences Amongst the Saucerians* (under a pseudonym, naturally) and passed it to an unscrupulous literary agent (not my current one, naturally) who in turn passed it on to an unscrupulous publisher (not my current one, naturally) the chances are fair that we would all make a fortune! The public has a staggering willingness to believe in the most exotic propositions. Witness the response to spiritualism, to religious 'manifestations', religious cults, ghosts, fairies and extra-sensory perception.[2] It is not necessary to either believe or dis-believe in any, all or none of these to appreciate how interested the public is in the exotic.

Like UFOs these phenomena are accompanied by the usual camp-following types of evidence—masses of reliable eye-witness reports, press coverage, elaborate theories and photographs which cannot be proved to be fakes. (Interestingly enough, the best example of photographic evidence applies not to UFOs or ghosts but to fairies.[3])

There would appear to be some form of psychological mechanism working in the mind of the human being which predisposes us not only to believe in the exotic but to interpret non-exotic happenings in exotic terms. A fireball, for example, could be seen as an angel, a UFO, a demon, a ghost (or spirit), a psychic or an extrasensory manifestation depending on which exotic mechanism was operating on or within the perceiving apparatus at the moment of perception.

One of the more interesting experiments performed to highlight this activity was done within the context of UFOs. Under hypnosis and monitored by a polygraph 'lie detector' some people who did not claim to have any experiences with UFOs 'truthfully' described in detail their meetings with beings from other worlds in strange spacecraft which had picked them up.[4] Should this prove to be the case generally then either most of us have been subjected to saucer scrutiny, the memory of which has been hypnotically suppressed by the naughty little ETIs, or there is indeed some exotic-oriented mechanism at work within the human mind. If the latter is so, then we find ourselves close by if not actually in the area of speculation regarding flying saucers which Carl Jung[5] suggests is the important one. Jung sees the circular and ellipsoidal lights as symbolic of a desire for 'wholeness', and maintains that this symbolism is consistent with that found in the world's religions.

This has been followed up by a number of psychologists who see in UFOs a subconscious attempt to fill the space created in Western Man's psyche by the destruction of religious belief. Interestingly, a great deal of the literature on UFOs, more particularly among the sensational 'down-market' works, concentrates on the superior nature of the ETIs, who are almost exclusively human in form. They are not only our technological betters, but our moral ones as well.

Something as fundamental as this within the nature of ourselves has bothered a number of those who have dwelt on the possible consequences of CETI. It bothers me, too. I am even bothered about those who are bothered about it. Let me explain. There is the 'Zealot' who wishes us to reject the whole idea of CETI and tells us not to answer the space 'phone when it starts ringing. Such individuals fear the consequences. Their fears are rationalized by reference to historical analogy but perhaps are really rooted in the exotic-orientation mechanism. Is their fear really rooted in the fear which we might all feel at the prospect of coming face to face with God? This is the obverse of the awe-wonder-and-love side of the exotic coin, and is every bit as much a 'gut' reaction as the other.

The possibility that such extreme reactions could cause chaos in the event of encounter is often borne in mind by those who wish to discuss encounter's consequences. There is some considerable bulk of opinion that, should the existence of ETI become an established fact, it should not be released to the public as the reaction could be outwith the ability of government to control. The old diehard illustration of this is Orson Welles' broadcast of *War of the Worlds* in 1938. The response to this was one of mass panic and it is seen as indicative of what could happen in the event of a real encounter. A friend of mine once said that news released of any kind of encounter would produce mass hysteria on a global scale.

One of the main problems these people see is the fact that so many of us already believe in the ETI and have a great deal of subconscious emotional capital invested there. I would venture a personal speculation here. An aspect of the flying-saucer days of the early 'fifties that tends to be overlooked by those who comb the newspapers of the time, particularly the American newspapers, for reports is this: what was happening on page one?

The atmosphere in these early formative years of UFO investigation by government agencies was a heavily charged political one. It is interesting to use it as a reference when starting the task of evaluating Projects Blue Book, Grudge and Sign. These took their investigative attitudes from an administration which grew increasingly paranoid as the Cold War chills ran up the spines of Americans both in and out of politics and which continued to shiver for over a decade.

My own feeling is that, in this atmosphere, concern over the possibilities of UFOs would flourish in particular ways. Firstly there would be the concern that there were little green men who posed some kind of extraterrestrial threat to the American nation. This is a fairly straightforward concern, but it has attendant problems. The second reaction arises from the first and is a dismissive attitude. Some people concerned with national security would be prepared to chuck the whole investigation into the scrap bucket if there were no results after the first eighteen months. As far as they were concerned, 'results' would mean a flying saucer brought down and being torn systematically apart in a military lab complex and the LGMs undergoing the third degree. The third reaction, and the one which I suspect is still operational to some extent, is a development of the second that took place in the paranoid hothouse of the 'fifties. Almost as soon as the dismissive reaction had become official thinking there would be some trepidation over the 'fact' that the agencies involved had been tricking themselves into believing that UFOs were possibly from Out There. *But what if it was not they who had been doing the tricking? What if it was ... oh no!* If the Commies were responsible for the trickery the effects could be catastrophic. After all, they had succeeded in pulling the extraterrestrial wool over the eyes of cool-headed security personnel. What could they then do to the dumbest of dumb animals, Joe Public?[6]

The interesting mental intricacies of the security-oriented mentality seem to act as a maze for ideas: once in they cannot escape but, at best, only wind up in different contexts. Even if the security forces became convinced that the KGB were not actively involved in propagating UFO hoaxes, the possibility that they may attempt this would have risen already. The basic fear which I can conceive them having is probably much less elaborate than those conceived by the professional national security paranoiac: could the Reds block communication channels in the USA by perpetrating a massive UFO hoax with the impact of the *War of the Worlds* radio broadcast?

Perhaps just thinking along these lines has distorted my usual optimistic thinking. As a consequence of the above I can see two simultaneous developments. The first is the official suppression of the idea that UFOs exist at all and the second, running as I say simultaneously, is the actual propagation of UFO reports (spurious, of course) to determine the public reaction.

Could the CIA have attempted a 'reciprocal' experiment behind the Iron Curtain?

Perhaps my reading of Miles Copeland[7] has left me a little too cynical. It still seems feasible to me that there will be tiresome consequences once the concept of an ETI encounter hoax having profoundly disruptive effects takes root. To the Intelligence-oriented, the possibilities of applying it to the enemy demand to be 'field tested'.

This is not the place to speculate on what Soviet UFO reports may or may not have been exercises (if they ever took place at all) in which Western Intelligence agencies tried out the principles of UFO hoaxing. Anyone who wishes to inquire in this area is not going to be given a great deal of help. Although some aspects of Soviet UFO reportage have been made available in the West, there is nothing like the requisite depth of domestic documentation now available in the USA, for example.

What the security agencies may be sitting so embarrassedly upon could prove to be the fact that they are indeed involved in ufology. But problems may be due first of all to the very fact that they do not wish dumb (but increasingly suspicious) Joe Public to realize that they are interested in discrediting UFO reports only for security reasons. I would be quite ready to accept as an explanation of certain UFO close encounters a statement detailing how they were set up as psychological test runs of public credulity by agencies responsible for national security. Not an easy thing to admit to taxpayers on whom the tests were run, who provide the funds from which the same agencies receive their appropriations. Some people, some in high places, too, would look very egg-faced indeed.

Problems would further arise if the supposition is true that UFO hoax has proved to be a credible psychological weapon. Personally, I am of the opinion that the security agencies are satisfied that it is valid in this area. For no really good reason, anything vaguely smacking of UFOs has been regarded as being associated with people who are publicity-seeking hoaxers, or slightly stupid inasmuch as they mistake clouds for spaceships, or mentally unstable cranks.

Interestingly, this attitude seems quite prevalent in the English-language countries. The curious aspect of this is that when we travel beyond these areas we frequently meet with attitudes which are strikingly different. In Britain or America, until fairly recently, no respectable journalist would have dared air a belief in UFOs. The idea that government ministers or serving members of the military would do this is still rather unlikely. In France, however, OVNI (as 'UFOs' translates into French) is a quite respectable topic for serious discussion. In 1977 the French Army's official magazine featured an article by one of the country's most widely respected TV journalists, Jean-Claude Bourret.[8] He reported on an increasing incidence of OVNI sightings. He quoted Colonel Gaston Alexis, M Robert Galley (Minister for the Armed Forces), M Cochard (Director of the National Gendarmerie) and astrophysicist Pierre

Guerin, a senior research member of the French National Centre for Space Research. A similar line-up of status-holding individuals by a status-holding individual being quoted on UFOs in an official military magazine in the English-speaking world would have been virtually impossible. Had it somehow been made possible, the attitude would have been absolutely the opposite of the French one: it would have been firmly dismissive of the phenomena as one unworthy of investigation.

At this point let me say again for the record, that UFOs *are* in my opinion worthy of investigation. Stating that many of them are simply plasmoids or temperature-inversion mirages is unsatisfactory. To be demonstrated as viable descriptions of the otherwise inexplicable phenomena these conditions should be understood, and they are neither of them well understood. Certain UFO sightings have all the hallmarks of a plasmoid or inversion-layer, but the theoretical conditions needed to satisfy applying them as explanations are not present. Investigation can only increase our knowledge of the world we live in.

Let me put another hypothesis forward here. There are, at the time of writing, reports of increasing numbers of sightings around airports (often military) and North Sea oil rigs.[9] I cannot help but speculate that perhaps the increasing communications traffic in the short-wavelength (high-frequency) areas of the electromagnetic spectrum may be associated with their appearance. Microwaves are effective in beaming energy from point A to point B, as has been demonstrated widely. Dr Glaser's proposals for shunting it down from orbital power stations in this form[10] and the increasingly common microwave oven in the kitchen are two examples. What effects would a freak focussing event have on the atmosphere around the focal point? Would normal but heavy traffic have particular energy interactions in as yet undefined atmospheric conditions? From a purely scientific point of view the subject seems worthy of investigation.

If it turns out that some UFOs are of ETI origin, whether spaceships flown by small green people or atmospheric probes sent out by larger robot interstellar probes, I will be the first to stand up and shout: 'I was wrong and being wrong's terrific!' Let me look foolish if genuine hoaxers have been at work; scepticism can be taken only so far. There is an immense popular following for writers of UFO books and for science fiction. The media's general cynicism for the subject may act as a buffer, but depending on how a hoax was perpetrated there may be many members of the public only too ready to believe and respond positively to it.

Whether or not agencies have quietly campaigned to make UFOs a subject fit only for ridicule, up until very recently ridicule was the official attitude. It was an attitude which tainted the whole of CETI in many eyes. Anyone taking a serious interest in advanced mental life elsewhere was regarded as a bit of a nut. This is probably why the early CETI conferences were not particularly interested in publicity.

In the event of an actual encounter, dismissiveness is, of course, the most dangerous attitude to hold. The shock which the event would have on not merely an unprepared psyche, but one which was antipathetic to the concept of ETI, might be, at least initially, quite damaging.

The problem is that UFOs and CETI seem to be synonymous in the public mind. This equation should be dispelled if for no other reason than that the attitude towards UFOs tends to be either uncritical acceptance of extraterrestrial origins or, as mentioned above, ridicule. A belief that UFOs come from outer space has a concomitant belief that intelligent life exists elsewhere, but a belief that intelligent life exists elsewhere does not imply an acceptance of UFOs as being extraterrestrial in origin.

In other words, the Viking results (which I interpret as indicative that life or something close to it exists on Mars[11]) do not mean that flying saucers come from the red planet. This kind of thinking, speculative but always critical, is the best frame of mind to adopt with regard to the whole subject of CETI. It is the kind of attitude which I and others interested in the problems which it creates often strive for. Unfortunately, in too much CETI thinking, probably my own included, there is a lack of imagination behind both the speculation and the criticism.

10

Them

One afternoon on glancing through my local newspaper I noticed that there were two science-fiction movies on in the one cinema complex, *Star Wars* and *Close Encounters of the Third Kind*. Being a movie gourmand, particularly an SF movie gourmand, I decided to spend the day gorging myself on both. With mind suitably cranked into 'gee-whiz' mode I settled down for a five-hour stint before the giant screen. A profusion of dazzling lights and giant-laser-blasting spaceships later, I emerged pleasantly stunned and grinning into the teeming streets of Friday-night Glasgow.

This is not the place for movie criticism and I am no student of the cinema, but I enjoyed those offerings immensely. Nor was I alone in this. The box office receipts for both films are now part of cinema legend. They were followed by a profusion of bandwagon-mounted products, few of which had either the daring or the originality of these two front-runners.

What was of particular interest was the attitude to ETI in the presentation of these fantasies. Neither of them made any great impression on me, although I was not disappointed. Both were very terrestrial in their conception. *Star Wars'* ETIs even knocked back the booze together in the local cantina. The behaviour of the *Close Encounters* creatures was slightly less common and even hinted at non-human reasoning, but sadly they turned out to be Little (Very Pale) Green Men.

The finest examples of ETIs in the cinema must be in *Solaris* and *2001: A Space Odyssey*. *Solaris* is in fact about the problems of men on another world with a sentient ocean. The two widely differing sentient lifeforms vainly attempt communication with each other. In *2001* the ETIs never appear, only their works and their power are in evidence. Both movies point to real possibilities. *Solaris* indicates that on other worlds evolution may have taken a different route from that adopted here on Earth; in this instance the profusion of organisms found on Earth is represented by the immense complexities within the structures of the ocean itself.

These movies are more than just good SF movies, they are examples of SF at its finest. When it is operating well, SF is philosophical fiction, and herein lies a great deal of its appeal. There has been much made of the escapism factor in SF, too much: there is more escapism in the average spy thriller because, *not* in spite, of the fact that it allows the reader/viewer to escape into a fantasy world within his or her own real world.

The amateur philosopher is generally unschooled in philosophy, believes it to be an obscure academic discipline and is unaware of the fact that he or she indulges in it frequently. Virtually all SF fans when they meet and discuss the subject compare concepts of time and space which, if not as sophisticated as those tackled by the professionals, are at least as wide-ranging.

The discussions which followed the first showing of *2001* in Glasgow were among the most stimulating I have heard. Unfortunately, the fact that *Solaris* was sub-titled and screened at our local arty cinema precluded the possibility of there being similar discussions. What emerged quite clearly was that there is a great number of people who are readily prepared to accept the possibility that other mental lifeforms can be even more different from us than we can presently imagine.

Perhaps this is why the concept of the amorphous being portrayed in Fred Hoyle's *The Black Cloud* became widely known in the field of CETI. As an example of SF it is first class and carries the seal of academic approval, having been written by an established scientist of impeccable, if controversial, reputation. How many scientists would have referred to it as it originally stood, an idea which was once tossed around by Thomas Gold? In the form of a novel it attained general currency. Nobody should think that this made it the province solely of the scientifically minded. No SF should ever be that.

In fact, it is possible to find examples of almost all the differing forms of possible biochemistry imaginable in the SF books and stories which have been turned out over the past few decades. In the works of one writer alone, James White, can be found a vast profusion of beings in a set of tales ('Sector General') based around an interstellar hospital.[1] These range from amoebic blobs, to insectiles, to chlorine breathers with weird and wonderful behaviour patterns.

The possible biochemistries for life elsewhere appear limited from our point of view, but then our point of view is limited by our knowledge of chemistry. We cannot state what all the reactions of all possible elements and compounds will be in relation to one another under every conceivable permutation of physical conditions. As a consequence of this, we tend to be, as Carl Sagan would put it, oxygen chauvinists. Let us take a brief glance at what a few of the alternative chemistries might be.

A great deal of water slops around inside us. We are virtually fish which migrated landwards, carrying with us our own oceans inside our skins. There is good reason for this. Water is an ionizing solvent: not only does it dissolve things, it also encourages them to form ions, which readily combine with other appropriate ions. Water also has a kind of handling mechanism in the form of loose hydrogen 'bonds': these hold the water together in strings and nets of molecules and are ever changing their configuration. This is useful. The loose bonds also set up conditions for biochemical activity to take place and are particularly important in aiding and abetting the genetic hocus pocus that goes on deep down in the cell.

That is what water does, but perhaps there are other equally useful properties in other liquids. Ammonia is one which both J. B. S. Haldane and V. A. Firsoff have considered. P. H. A. Sneath[2] mentions the possibility of still more, such as hydrogen fluoride, hydrogen sulphide, and even hydrogen cyanide. Creatures developed within the chemistries of these compounds would, of course, regard Earth as a highly unlikely abode for living beings, it being strongly polluted by that highly inflammable, corrosive and poisonous substance oxygen!

Carbon is one of the basic building-blocks of all known life, but here again there is the possibility of another approach which would be successful for life attempting to develop in a carbon-poor environment. There are substitutes or at least analogues for carbon to be found in the table of elements. Silicon, for example, is a particular favourite of the SF writer: sluggish stone-like silicate beings scrunching away at rocks and excreting gravel have been making their lugubrious ways through SF since the days of Stanley G. Weinbaum's pyramid creature in his 1934 classic *A Martian Odyssey*. It has now gone out of fashion as an alternative chemistry but perhaps new knowledge will bring new possibilities, as it could to germanium, another analogue of carbon.

At very cool temperatures the so called 'noble' gases, which are generally fairly inert, actually form compounds. In contrast, sulphur might act as a vital solvent at really high temperatures as Hal

Clement suggested in his novel *Iceworld* (the title planet being Earth).

Given a planetary environment which is different from that with which we are so very familiar we should expect that, across the thousands of millions of years of the planet's existence, there are liable to be some chemical peculiarities arising. We should not assume that these will necessarily lead to life of any sort but they may lead to some very exotic conditions.

There may be no life on Mars, for example. The results from the Viking landers' biological labs proved to be ambiguous. If there is no life, then the chemistry of the Martian soil is odd indeed. This in no way contradicts what we have learned about the Universe over the past few millennia. We have come to expect the unexpected. In particular, we have over the past century come to regard the apparently inexplicable or paradoxical as being the norm. A problem is only a real problem when its solution creates further problems.

We must expect bizarre chemical and biochemical activities to be present in other parts of the Universe. By their very nature we cannot confidently predict what they are likely to be. As a consequence of this, we should be very wary of predicting what course of increasing complexity evolves therefrom. Using the rise of life on Earth as a stereotype is all very well if you are painting your picture with a really broad brush on a very coarse canvas.

What is really not viable is any attempt at drawing analogies between the rise of intelligent life on this planet and its putative rise elsewhere. It is fair to say that the evidence for chemical evolution indicates that complex chemical phenomena may produce chemical systems elsewhere, which, if we could examine them, we would eventually decide to describe as *bio*chemical. How the life-forms would change and develop from that point forth we are in no position to say, but we do have an indicator as to one factor— it will almost certainly be in many respects profoundly different from anything which we have come to imagine.

There has been a fair amount of very penetrating inquiry into both the nature and the evolution of intelligence. Some writers and CETI specialists freely apply this to the 'Extraterrestrial Civilizations' to which they frequently refer. I must reiterate that I find this to be the most staggering conceit anywhere in the numerous volumes devoted to the topic. My own suspicion is that, if we are ever to come to terms with an ETI, we must regard the concept of intelligence as a very plastic one.

What then can we say about advanced mental life in SETI/CETI terms? First of all, by virtue of the fact that it is alive it is an information store; secondly, it uses its information in such a way that it adds to it; thirdly, it has as its basic expression of that information a relationship between itself and its environment; fourthly, and finally, an expression of that relationship is an interstellar urge (or we wouldn't be encountering it).

The axiomatic bases for an AML's mathematics may therefore be not merely different from our own, but the being may not even be aware of the functioning of its 'mathematical' faculty in the same way that we are. Processes like logic and mathematics may run at a subconscious level for beings with a powerful, wide-ranging computational faculty at the very foundations of their mental development; concepts such as zero and infinity might mean little to them, particularly if the physical configuration which gives them this faculty is distinctly different from the human brain.

Samuel Aronoff has pointed out that social evolution is the partitioning of information and response processes within a species. In the case of a non-terrestrial AML, the selective pressures on how such partitioning comes about are liable to be different in a number of respects from those on Earth, which have been operating generally within the history of Earth's environment. This very probably being the case, what we must look to is not social evolution among our fellow inhabitants of the Universe but a great variety of mutually exotic processes each to some extent analogues of the other.

Joseph Royce, who has devoted a great deal of his professional life as theoretical psychologist to the proposition that we, as human individuals, are encapsulated within a particular way of knowing, believes that there could exist alien 'world views' which would have a tremendous impact upon our own.[3] He sees us as limited, partly genetically, to epistemic encapsulation. Our consciousness is a product of three factors: reason, symbolism and perception. R (Reason) + S (Symbolism) + P (Perception) = Human consciousness. The mechanisms which support these factors may differ from AML to AML. Indeed, the factors which dominate the mentality of an AML may not even be closely allied to any of the three which are essential to our own. In another AML there could be at least one X factor which could combine with or substitute for R, S or P in ourselves, or even supplant them entirely, which would give us combinations such as $RSPX$, RSX, RXP, XXP, SXX or even ultimately XXX. The latter would be by far the most difficult beings to face intellectually. Imagine something which we saw as a really sophisticated 'technological culture', a 'being' which had totally alien ways of perceiving the Universe, did not interrelate symbols in the way which we do, and had fundamentally different bases for its thought processes particularly in areas we explore with logic and mathematics. If, on top of this, it was principally involved in intraspecific communication by something akin to telepathy, we would be in a position where contact of a meaningful nature was virtually impossible.

We should come to the conclusion now that there are possibly more interpretations of the Universe than merely the human one. Beings evolving differently (we may not even view their developmental history as true evolution in our terms, in fact) are likely to have a separate 'angular truth' from ourselves.[4] Try to envisage advanced mental life, including ourselves and the cetaceans on planet Earth, as having mentalities shaped like tennis rackets without handles. They are limited in their physical structures. They have differing 'diameters' and 'mesh sizes' and are arrayed at different angles to one another. To take this wild image into wilder territory you must imagine the Universe around them as a torrent of tennis balls all zooming in different directions, all travelling at different speeds and no two of the same size. Any racket can interact with any tennis ball travelling in such a manner that it will come within the boundaries of the racket. The problems arise where, in the first place, the balls are smaller than the mesh size, in which case many of them are liable to pass through without reacting —some small ones will, of course, hit the mesh as a matter of probability. The second problem is with balls which are travelling at a very shallow angle to the racket, as these are likely to zoom past (as are those coming in at a steep angle but outwith the dimensions of the racket).

Holding this rather strained figurative description of AML in the imagination, we can try to express SETI and CETI as attempts to draw together two rackets. If we succeed, the difference between ourselves and the other AML will be principally angular: the areas of clearest correspondence will be concerned with the tennis balls which interact with both of us at the same angle. The greater the difference in angle the more difficult becomes the possibility of exchange, but the greater becomes the possibility of breaking out of our 'epistemic capsule'.

If the analogy holds true then it should be possible for both ourselves and non-human AMLs to enrich our views of the Universe, including our views of ourselves. This enrichment may well lead to new methods of approach to the problems of CETI and perhaps even to the creation of translator beings, whose mental structures can encompass the singularities of the two 'parent AMLs'.

Should this ever come about, we may find ourselves having to contend with areas which we literally do not have the requisite equipment to tackle properly. To examine this means plunging into the area of the unbelievably fatuous hypothesis. Well, with fair warning given, I am about to splash into these waters.

Let me examine a purely speculative hypothesis which I am not going to put forward as anything other than fantasy. It is rightfully placed within these pages only as an illustration of how different our abilities may be from those other creatures out in the Galaxy.

I want you to consider the possibility that supernatural phenomena are attempts at CETI by beings with a very different mental structure from our own. This accordingly has different functions, the principal one of which interacts with the Universe in what we would describe as psychic terms. In this scenario ghost phenomena would be 'replays' of events picked up by the ETIs, and rebroadcast to us in much the same way as some scientists imagine that the first signs of radio contact from another star system might be us hearing replays of 1930s radio programmes, beamed back at us to catch our attention.

Let us assume that there really is a psychic energy phenomenon associated with all living beings. Then we should also assume that that energy can be transmitted in some manner which the ETI can appreciate. Perhaps the moment of death discharges a great deal of information in this psychic form from the dying organism, either directly about itself or about certain events in its life. What happens on rebroadcast is that information about the dead reappears in a variety of manifestations.

If this is the case, then either the psychical tennis balls interact with our mental tennis racket at an extremely shallow angle, so that we seldom strike them or only at the rim, or they are too small for our mesh to field consistently. In either case, we are partly aware that something is happening but because it is so inconsistent with our own mentality we cannot accommodate it.

This leads to a number of interesting speculations. Hypnotic subjects regressed beyond their own lifetimes have been known to produce evidence of 'previous lives'. Could reincarnation be psychical CETI? Likewise, could the old stories of heavens and hells merely have been crude attempts at conveying extraterrestrial environments to the human mind? Likewise demonic possession, vampirism and lycanthropy might be memories of ETIs coming across strongly, into what they did not realize was an environment which could handle the psychic aspects of their minds only minimally. Furthermore, concentrated psychic power, making use of psychic communication masers or lasers, being received by human beings who were as optimally tuned as was possible for a mere human would give those individuals apparently supernatural powers.

And, of course, if the psychic powers were operating outwith Einsteinian space-time they could provide psychical data about the future . . .

Hopefully the general picture comes across from that reckless flight of fancy. It may be that ETIs are indeed trying to contact us but that we cannot appreciate their attempts for physical/technological or mental limitations on our part. To be realistic about the psychic potential we might say that a telepathic AML might attempt to embed images in our consciousness which it felt we would associate with CETI. Who knows, could this account for the numerous people who have under hypnosis proved to hold encounters with UFO inhabitants locked away in their subconscious minds unbeknownst to themselves?

To belabour the tennis-racket image, I want to point out that the whole picture is of billions of rackets—all shapes and sizes and with various mesh patterns and dimensions—arrayed in a random scatter of angles. Nor does it stop there. The picture is not static. It moves. The angles change at different rates as do the sizes and shapes of the rackets and their various meshes. New rackets continually appear among the old and all of them, like snowflakes, are quite singular: each is perfectly unique, each an exotic treasure of mysterious mentation.

High Society

Many voices before mine have said that we are the only civilization in the Galaxy. I add mine to the chorus but with a descant. In terms of what we conceive to be intelligence we may in fact be virtually alone in the Galaxy—if not the Universe. But in terms of powerful and profound mental life I at least feel certain that we are, as they say, not alone.

Enrico Fermi's paradox 'Where is everybody?'[1] is resolvable by demonstrating that there is no paradox. If AML has evolved over the billions of years of Galactic history at a fairly regular pace where are all the ETIs? If they are all different not merely from ourselves but from one another as mental lifeforms then we must define clearly what we are referring to in the 'body' part of 'everybody'. Are we not saying in effect: 'Where are all the spaceships, the colonists and evidences of interstellar exploration, conquest and empire?' Is this in effect our statement: 'Where are all the beings which think like us, have our picture of the Universe and our motivational drives?' There is therefore no paradox. They are all around us. We are 'everybody'.

For an AML of non-terrestrial origins the motives lying behind any interstellar urge which it may possess could be so different from ours that they could be literally inconceivable to human beings. Concepts such as 'conquest' and 'empire' are quite distinct from 'exploration' and 'knowledge', but both sets have a similar relationship and similar 'roots'—terrestrial life.

Should we ever encounter a radio CETI attempt from elsewhere or discover an extra-solar spaceprobe in our system we should accordingly be careful in our interpretation of the ETI's motives. Our interpretation of them might safely be that there is a principle of extension at work within the behaviour of the ETI. We cannot say other than that *within our terms of reference* it is probably exploratory.

How much advanced mental life is there likely to be out there anyway? The estimates vary comically so there is no point in talking about a count. Let us assume that we are very much the average. In this case, the average time for a given body of stars to produce an AML is our own five billion years. This means that half the number of AMLs which a group of stars is going to produce have already appeared at the end of five billion years. Taking a body of stars to be analogous to a chunk of radioactive metal which pings out particles the way the stars ping out AMLs I am going to guess-timate that any body of stars has an AML 'halflife' of five billion years.

If this is so, it is fairly obvious that most of the Galaxy's present AML potential has been used up. This does, of course, depend on the rate of star formation; accordingly I am going to consider only the stars as old as or older than Sol. The first consignment which arrived some fifteen billion years ago should have now produced almost 90% of their AMLs, the middle-aged group a mere ten billion years old should have given rise to about 75%, and we are at 50%. Thus if we are average then the Galaxy to date has produced over 70% of its AMLs and we are arriving late in the race.

Let me now tackle star formation. It would appear that stars probably appeared more rapidly during the earlier eras of the Galaxy than at present. To balance this up, the stars now popping out of the interstellar nebulae tend to be richer in complex heavy elements. I will assume that the balance here is approximately justified for a fairly steady rate of AML appearance. This means that the *percentage* of produced AMLs is continuing to rise.

Who knows what the real numbers are? They may be billions or a few dozen but the assumption of mediocrity tends to put us in the picture when most of it has already been painted. This is no bad thing. Assuming that some form of CETI has been established (probably quite different from any we are currently fooling around with in the kindergarten), there is a likelihood that the Galactic Club of Ronald Bracewell[2] really does exist. An intercommunicating community of developing mental lifeforms may already be thriving in the stars around us.

There may in fact be a kind of reverse zoo hypothesis (the zoo hypothesis suggests that ETIs are watching us much as people watch animals in a zoo) whereby we have to wind our way down a tortuous intellectual, psychological and technological maze before we can even make ourselves known to these ETIs as beings significant in their own terms. This really will make it the toughest of all possible clubs to penetrate.

What happens then to the lifeforms who cannot make themselves heard no matter how hard they wallop the door? Do they form another club? Could there be a variety of clubs coexisting in the Galaxy within the same regions of space which are only dimly aware of each others' presence, if at all?

To gain entry to an exclusive club one is generally introduced by an established member. No amount of ringing the doorbell alone will suffice. Accordingly, what we must search out among the stars is an AML which has 'racket' characteristics as close to our own as possible. This may prove to be quite a task. We may well find ourselves to all intents and purposes alone in the Galaxy for centuries if not millennia (although I have aired my doubts about the survival of us as *Homo sapiens* for more than a few more hundred years or so).

On reading Carl Sagan's *The Cosmic Connection*,[3] I was rather startled to find that the eminent scientist seems to regard the formation of a Galactic Club as a process of homogenization. Quite apart from the fact that I view this prospect as appalling, I can hardly regard it as being credible. Even if the various AMLs in the Galaxy are not as disparate in nature as I believe them to be, the postulate that homogenization can take place as a Galactic phenomenon is certainly slightly cross-eyed. Acculturation is a two-way process: to begin with, both or all of the interacting cultures change in response to one another; secondly, they continue to change in response to the responses of one another. On a small scale with good communication this can result in homogenization, but the conditions must be kept stable.

To start small and work upwards I will begin in childhood. I was raised in a tenement with a dozen families living up the stairwell (or 'close', as we call it in Scotland). We had Catholics, Protestants and a Jewish family. Sometimes we all lived in stupefying harmony, occasionally it was bedlam, but always communications were first-rate. Heterogeneity as much as homogeneity was encouraged by this set-up. It was very easy to acquire some aspects of your neighbours' behaviour. It was equally easy to eschew that which you found distasteful. To this day matzohs do nothing for me but gefilte fish are fine. The result was an ever-fluid heterogeneity.

The point was that, although we lived in a densely crowded local community, we had many channels of cultural or even subcultural access outwith it, as we lived in a city.

Damascus is the oldest surviving city on Earth. In spite of the fact that it has been around for some thirty-five centuries there is no sign that it is any more homogeneous a community than any other city of similar size in the world today. Furthermore, cities have always been loose amalgams of many differing kinds of people. They may have language and religion in common, but the variety of activities which go on in the myriad lives during a single day is vast, the personalities never duplicated, the opinions frequently conflicting and family customs and traditions differing from house to house. The city is too large for homogeneity.

The rural community does tend to homogeneity because there is comparatively such a small number of families in a small geographical area with little access to activity further afield. Interestingly enough, when you spread the geographical area to encompass several rural communities an element of the heterogeneous seems to emerge. Activities in one community will differ sufficiently from those in another for the inhabitants of both to be aware of them. There is an old Glasgow saying that 'they're queer fowk who live in the Shaws'; in other words, the people on the other side of the mountain are always a shade peculiar—they really don't do things the way they *should* be done.

On the global scale it is easy to mistake the ubiquity of Coca-Cola for a growing process of homogeneity of intercontinental proportions. This is so of most modern cities. Battery-human architecture permeates the metropolis from Djakarta to Brussels. There are transistor radios in Kinshasa and TV sets in Ulan Bator. My own reponse to this is that they will probably have the same impact on heterogeneity as did the wheel: they will enhance it. Admittedly, the wheel has been trundling around somewhat longer but the impact of cheap mass-produced electronics on communication is likely to be just as (excuse the pun) revolutionary. This means that what we are liable to find is not the Global Village,[4] not an expanded version of a rural hamlet, but rather an extension of the city. Megalopolis is indeed likely to cover the world, at least figuratively, within the next fifty years.

There should therefore be an increase rather than a decrease in heterogeneity.

Where something new appears within the ambit of an individual's consciousness he or she can accept, reject or simply ignore it. For example, let us remember that five hundred million people around the world watched a country boy from Wapakoneta, Ohio, make one giant leap for the United States when he placed his boot on the

Moon. Thereafter the television audiences dropped dramatically, picking up only slightly when Apollo 13 promised, and lived up to, the space fiction papped out by numerous TV stations around the world.

There was no homogenization. The world did not overnight accept or reject or even ignore spaceflight. What tends to happen in the late twentieth century is that individuals having access to such a variety of communication sources produce a variety of personal opinions and responses which become in themselves new sources for others. Even where there is a quite distinct argument between two factions on any subject, there tends to be a wide and changing span of opinion on both sides.

There is a school of thought that would have me think that the very heterogeneity found increasingly around the world is but an indication of the much more subtle process of global homogeneity. Travellers and tourists complain that foreign parts are losing their regional flavour. This is partly true but not particularly important. There is still and has always been a great preponderance of local preference and tradition which novelty does not destroy. Preference and tradition are, however, not nearly so static as some people seem to believe. They change. Increased communication facilities will help them change more easily. Nevertheless, I feel that, despite the appearance of Chinese and Indian restaurants in the East End of my home town Glasgow, the Big Dirty City will never become a blend of Delhi and Hong Kong sublimated in a universal Manhattan-type gel.

Societies generally, as well as cities, seem to be becoming increasingly heterogeneous. There is a greater diversity of factors operating on the lives of individuals in most parts of the world than in previous times. Admittedly, poverty is still the dominant factor for the majority of the human race, but even among the poorest new influences are bearing in upon their lives, although in their case seldom for the better.

Globally we are aware that fashions in dress as well as politics can flit rapidly around the world. Even within the confines of our planet we are becoming increasingly heterogeneous. When we move into the realms of interstellar exchange we take another quantitative leap. If I am wrong and communication is by means of long-distance radio messages, we will never listen to the same star twice, to paraphrase Heraclitus. Human culture is liable to change in response to an ETI's signal and the ETI is itself going to change in response to our signals if it is interested in communication.

It would be foolhardy as I have mentioned earlier to regard other AMLs as having the same kind of social structure as ourselves or anything remotely similar (they may be an 'it', a solitary vast and complex organism as in *Solaris*). To extend this line of thought, I have to confess that I regard the idea of a pan-Galactic super-civilization as an overwhelmingly anthropomorphic one. To that mutually agreeing group of CETI thinkers who see the Galactic Club as some kind of super United Nations I must quote the words of Sam Goldwyn, 'Include me out!'

'There is no point in simply demolishing something unless you have an alternative to put in its place.' I agree. I attacked the current feeling that we should listen for interstellar radio signals, preferring an investigation of Earth and the Solar System for extraterrestrial artefacts. Likewise I would like to put up an alternative type of Galactic Club to the supercivilization so beloved of my fellow SF writers (Galactic Empire is one of the cherished images of many very good SF writers) and of some CETI specialists.

We may be alone in the Universe in that we are a unique AML. We may go exploring the planets and move out into the rich star-fields far beyond and find nothing remotely similar to Man's mind for a long time, centuries or perhaps even millennia. We may be aware that there probably is a great deal of exchange taking place between lifeforms with which we simply cannot communicate.

There are two strategies open to us. The first is to go on search-ing until we come across an AML which can at least partially interact with our mentality. The second is to try kicking in the door. The latter would amount to an ultimatum which hopefully would force the ETIs to attempt contact with us instead of coolly ignoring our presence. Let's say we grow up technologically as nasty as we are today and decide to visit the home system of one particularly supercilious bunch of jumped-up brachiopods and slowly set their star on the road to going nova: 'Speak to us or we'll blow up your sun, so there!' I will not dwell on what the consequences of action like this is likely to be. I prefer to imagine that by that time we as a species will be considerably more mature than we are presently. If it looks as if we are not going to mature then we should take steps to bring this about. It would be irresponsible to let Man loose upon the cosmos, as some kind of superpotent technological vandal.

It may be necessary to manipulate our genetic material in order to produce a really sane species. Koestler and others have argued that there is something profoundly pathological in the very nature of Man.[5] We may have to change our nature and in doing so become something other than Man. Perhaps a new gland may be introduced to our physiology which releases an enzyme particularly designed to suppress those physio-psychological mechanisms which encourage us to indulge in such sports as avarice, egotism and war. The presence of other enzymes which cool the body temperature but stimulate all mechanisms so that they operate at normal body-temperature rate may be the answer here. The structures we wished to suppress would be provided with a barrier to these enzymes and would thus function at a low metabolic rate; in other words they would be sluggish compared to the rest of our activity. We are certain to tumble wildly down many well disguised pitfalls in exploring territory such as this but, to play with the grown-ups we will have to give up behavioural short pants and tailor ourselves some decent adult clothes.

The strategy of finding someone who can listen among the ETIs is the only viable one for entry to the Club. Searching among the stars may produce a likely sponsor, one who can equate a certain 'angular' appreciation of the Universe with one of our own.

To return to the reverse zoo hypothesis, where the ETIs are there in open evidence before our eyes but inaccessible, let us imagine that there is a party going on inside a large house. We suspect that a lot of our friends are likely to be inside enjoying themselves and we resent being left outside in the cold lonely winter night. No amount of ringing the doorbell is going to help. Everyone is having too good a time to bother answering it.

The only solution is to pick the lock and walk in.

If there is no way of finding someone to talk to us in terms which we can understand then we must find a way of talking to them. Mohammed is reputed to have had a similar problem with a mountain. In a way this resembles the choice between listening for radio signals over the vastness of interstellar space and actively scouring the local interplanetary neighbourhood for renewable spaceprobes. Instead of hunting for something which may or may not be there, in this case an ETI which can think like us in some respects, we should develop a machine which is capable of producing non-humanistic views of the Universe based on data about the lifeforms concerned. (Here the word 'machine' is being used very loosely. I am assuming that when this event takes place, when we are preparing for membership of the Club, the line distinguishing between the human being and his creations will have smudged virtually out of existence.) Earlier in these pages I speculated about there being colleges of vN in most star systems which would have most likely evolved their own *lingua Galactica*, and this would certainly act as a starting pointer for us.

The problem which we shall face after encountering the vNs is the fact that although they will represent a 'frozen' common level of what we see as technology, the AMLs which created them will have continued to change. This is important because of the disparity between astronomical and human timescales. A few thousands or millions of years is, astronomically speaking, quite insignificant but to us is a span widening from the history of human culture to the very appearance of mankind on this world. If vNs have been arriving with some regularity over the past five billion or so years we have, as you can see, something of a daunting task on our hands. The Advanced Mental Lifeforms may be just so advanced that they are totally beyond us.

If this is so for us then it was probably so for each AML on reaching the vN equivalent stage. In this event there may not be a Galactic Club for us. Perhaps all ETIs exist in complete temporal isolation from one another until such times as their 'cultural' histories begin to take on astronomical proportions and the age differences grow less meaningful. This would have an analogy in human terms of how important an age difference of, say, ten years is to two individuals. When one is twelve and the other two years old the difference is enormous. When the former is ninety-five and the latter eighty-five the difference has become almost insignificant.

In this sense, entry to the party is a parallel to being born. But the onus of childbirth, all the pain and struggle, belongs to ourselves. The rewards which lie in that new life are completely outwith our present grasp.

We must change to be 'born' in the same way that the fertilized egg changes to embryo, and the embryo to foetus, before birth can successfully take place. In our case the physical changes will be the least important except inasmuch as they allow increased enhancement of our minds' capabilities. With the blending of Man and machine physically and mentally the synergetic conference of brains functioning as superminds could become a reality.[6]

Bernal envisioned a society in which human brains were interlinked, supported and enhanced artificially. This is a possibility which should not be ignored. The problem with organic material is death. What may well happen is that the intricate and dynamic correlation of electro-biochemical elements which comprize the brain could be transcribed into crystal where the lattices and interstices might take the part of cells and synapses in our nervous system (see page 78).

At this point we would be able to have almost total control over our further evolution. We would be able to produce gigantic crystals containing millions of minds in chorus, all extended in their capabilities and enhanced within those capabilities by integration with artificial intelligence. At this stage there really will be no point in trying to differentiate between Man and his machines. At this stage we will have an alternative to death, because, apart from the possibility that the crystals will be nearly immortal, they will be capable of transferring memory and relationship structures. They will be capable of remembering how these can be regenerated, which would amount to a cloning of minds. There would also be the potential to produce deliberately a far greater mental diversity among the new species than there is among our own. Diversity is the keystone of survival, so this 'solid-state society' would have considerable survival characteristics.

If the *lingua Galactica* of the vNs exists, it should in some instances be able to help us set up a new science of environmental behaviour studies, which we call ethology but which would in this context become xenethology. With luck some vNs will contain details of the developmental history of their makers. This will give us a lead, however tenuous, to the motivational structures within the AML which built this probe's ancestor in a remote part of the Galaxy in the very distant past. We should endeavour to observe a number of such cases, illustrations of AMLs going through processes akin to our own physical and behavioural evolution from a primitive single-celled bacterium to a complex mammal with involved social graces and disgraces. This could illustrate how the various angular differences between mental lifeforms may orient in regard to one another. We may have to come to terms with different developmental processes, some of which may not even be particular closely allied to evolution in the terrestrial sense at all. Being an optimist, I assume that if other AMLs have overcome this conceptual problem then so shall we.

We will then travel into the realms of abstract behaviour, the xenethological equivalent of mathematics, moving into the realms of 'imaginary or complex numbers', because we will be dealing with the translation of non-human behaviour into terms which we humans can understand. There will be only certain aspects initially, and those the ones closest in relation to our own behaviour, which come across strongly. We will then be able to redesign and create new minds which can cope more readily with both the human and non-human. We will be creating differentials.

When we are really adept at assembling the differential intelligences we will have altered the conceptual tennis racket with which we swat the Universe's tennis balls. The mesh size will be that little bit finer; the racket face will be beginning to curve and grow wider. The change will be significant. When it becomes sufficiently significant for us to set up meaningful interactions between ourselves and one of the Club we will have arrived on the other side of the threshold, stepped inside the door, moved into the new world.

Diverse flexible AMLs may thus set up mental differentials between themselves and other AMLs. It will be the equivalent of communicating with one entity with whom we have nothing in common *via* another who has something in common with us both. These others with whom we can communicate to some extent, the Human Analogues, will dominate our relationship with all members of the Club. Those areas within which we have achieved a common angular perspective will become increasingly important not merely to ourselves but also to them.

Now we come to the point which is perhaps the most hubristic of all. We will be prized and valued members of the Galactic Club. If, as I suspect, all forms of Advanced Mental Life are mutually exotic, then each will be prized by the others as a new and enlightening approach to the problems of existence within a Universe which can be understood from so many different and seemingly inconsistent viewpoints. We will not only require differential AMLs for communication but we will act as differentials ourselves for others, enriching the ever-changing dynamic of the Galactic Club.

From this point on there will continue to be great diversity and continued change resulting from it, but terrestrial evolution will have been superseded. We will no longer change through the fissioning process of random mutation and genetic drift alone: instead there will be a process of synthesis based on the mutations.

The new will be formed by creating deliberate as well as random variants and combining these to form still other variants. Thus, when we enter the Club, we will become a member in the most intimate sense. We will combine with many of the Human Analogues to create wholly original mental lifeforms.

As I have said, the Galactic Club will be nothing like an interstellar United Nations. What then might it be like?

I imagine it as something like Babel; an enormous conference of linguists each with a different behavioural language and each attempting to communicate with the other, the principal problem being that when they do make a breakthrough it changes their own language and the language which they have 'penetrated'. Some languages are impenetrable to certain others, and these need professional translators skilled in the languages which are mutually exclusive. For meaningful communication between all members, chain translations have to continue. The problem is that every message received changes the receiving language, and consequently its response is a novel variant which creates more change.

Special translators will have been bred specially to meet the needs of the conference. They are immensely flexible and capable of handling rapidly shifting diversities. They are the result of exceptionally close intercourse between conference members. The special translators are always at a premium because new members are always strolling into the Club with naïve misconceptions about everyone inside speaking their native tongue. These kids have to be handled delicately because their own language can describe aspects of existence which none other can. So we may expect a special translator as the Club's doorman, a staff member. I hesitate to speculate on the Club's executive committee.

An ever-changing population of vNs in the Galaxy might be both door and doorbell, elements with which we forcibly interact in order to let the doorman know that we are outside.

To appreciate how the Club might function at levels above that of basic membership, the Club analogy is best dropped. My own interpretation is that it might be better to employ the living body as a metaphor. The variety of differing AMLs might be regarded as the numerous teeming microscopic constituents of the single cell. Then there are the specializations of cells into various types of tissues. The tissues are employed by different systems working in close coordination to support the organism as an integrated dynamic whole. Here the vNs would be regarded as the basic material of the epidermis, the outer part of which is virtually dead. The live tissue lies just beyond.

For what it is worth, that is my personal vision of the Galaxy and Man's future within it. We are unique and as such highly desirable as a constituent of the vast heterogeneous system out there.

Perhaps, if we never achieve anything more than planetary devastation, we shall not be missed. The vanished potential will never have been appreciated. Let us then look to the stars as we have for millennium upon millennium and realize that the only route which can take us there lies buried in the dark of the human heart, that there is no way to reach it other than that of compassion and the will to change.

Epilogue

What Do You Say Before You Say Hello ?

The President's scientific advisors were really worried. The Press Secretary could see that. Generally they wore the standard patina of intellectual profundity with the care other career professionals lavished on the selection of neckties and office furnishings.

Today they sat, slouched, massaging tired faces with their soft hands. Nobody was name-dropping, 'seminal paper'-dropping or even conference-dropping. There was a lot of trouble somewhere and nobody was telling the Press Secretary a damned thing about it.

He scowled.

The President had told him to leave everything and come up to the Oval Room. This meant that within the next hour or two he was going to be under a barrage of demands from every paper between the *Herald-Tribune* and the *Bengal Engineers' Daily Digest*.

Everybody always found out before the damned Press Secretary!

The door opened and heads turned. The Man's personal aide, a balding youngster in his mid-twenties, twitched a smile at them from his thin face and indicated with a brisk semaphore of his right forefinger that they were to go inside.

'I don't want everyone to start speaking at once,' said the small cracked voice from the big svelt Philadelphian. 'Jess, you are the one who is really important in this room. You and only you ask the questions.'

The Press Secretary felt ice crystallize along the convolutions of his larger intestine. If he was the kingpin at this meeting the world was standing on its skull.

The lights dimmed a little and a television plate hanging from the wall splashed a shudder of colour through the room and then congealed into a picture.

There was a shallow grey dish-shape suspended there, and beside it, casting a shadow, was a large flat disc. Behind them, below them, was the unmistakable crusted acne of lunar Farside. Jess knew what this was. The NASA/ESA publicity machinery had been crowing about their miraculous radio telescope since it was completed three years back at the turn of the century. It hung in a continually computer corrected orbital alignment which held it forever some 60,000 kilometres above the other side of the Moon. Here it was completely shadowed from Earth's radio noise and annoying discharges from the planet's radiation belts. The great disc performed the same shielding job with regard to the Sun. The instrument was, apparently, capable of identifying radio binaries in the Andromeda Galaxy. He knew that was supposed to be impressive!

A clipped voice issued a hurried accompaniment.

'Two months ago the ten-thousand-metre dish above Farside picked up an unusual source during a standard calibration run. The source was issuing a complex repeated signal interrupted by what appeared to be other non-repeated signals containing elements related mathematically to one another and to the repeats.'

The picture became a bright blue pulse tracing against a black grid. There was a series of numbers ever rising in value at the top right beside a designation which read ETI PROBABILITY PROFILE.

Jess understood now.

'Deep space planetary monitors around Jupiter and Uranus obtained and confirmed infrared fixes placing the source at some five and a half thousand billion kilometres from us. Parallax coordinates substantiated Doppler information to the effect that this source is moving rapidly on a course taking it into the Solar System and the vicinity of the inner planets. On discovery it was travelling at 12% lightspeed. Its deceleration, if maintained, will place it in an elliptical orbit between Mercury and Venus less than ten months from today. Any doubts that this object was other than an extraterrestrial artefact were dispelled at 0113 hours Eastern Standard Time today when a reply came to the CETI experimental package automatically authorized, designed and transmitted by the ten-thousand-metre's twin IBM Fifty-Fiftys three weeks ago.'

A few whispered oaths assailed the ears of the unbelieving Jess. A message to aliens authorized, designed and transmitted by a batch of computers! What goes on?

The picture snapped to a grainy representation in monochrome. It was so long since Jess had looked at black-and-white TV that he had to do a double take. He saw a slowly turning ring with a metallic tube at its central axis. A firestream issued from the tube.

Some kind of rocket, Jess surmised.

'*It is now fairly clear that this is a world ship. The ring in the picture is hollow and contains a diverse rich biosphere representative of the ETIs' home world. Their power unit is under analysis and seems to be basically some form of mass annihilation process with which we are not presently familiar.*'

The screen was still showing scenes and talking when the Press Secretary stood up and for the first time in his life bawled at his boss.

'You allowed this bunch of shmucks to incorporate CETI programs into those Fifty-Fifties. You became so blasé about our contacting life from elsewhere that you let them philosophize happily away while the computers did the real work! Well, now we know. Now the whole bloody world knows!

'What do we do about it, Mr President?'

In the adjoining room a 'phone began to bleat.

This is the kind of science fiction which might become horrifying fact. So many repetitive tasks are digested by attitudes of complacency that it is often difficult to ensure that a great deal of control over them stays directly in human hands. It is often not merely tempting but necessary to delegate much of the work to computers. Where the task is interesting in some element it might be continued, but with results appearing no nearer more and more of the work is likely to be unloaded on the machines, especially as they themselves grow increasingly smarter. (To my knowledge there are no plans by IBM to produce Fifty-Fifty series computers, and so far it looks unlikely that we will see the ten-thousand-metre radio 'scope within the next thirty years.)

The ETIs of science fiction are often beings which, if not at our own level of technology, are a mere thousand or so years ahead. Personally I don't see this as possible, although even a gap of a few hundred years would be staggering. In the scenario above the aliens are perhaps one and a half centuries ahead. However, it may be feasible, with a stretch of the imagination.

There is a current theory that planetary systems are formed among groups of stars by a supernova[1] occurring within or near to that group. This would entail that many of the planets of nearby stars would be approximately as old as those of our own Solar System. Again I am not too happy with this. There is the old comparison with the age of the Earth and the eight thousand or so years men have had civilizations recognizable as such. Take the age of the Earth as equivalent to one day and civilizations have been around for less than the last one-fifth of a second. Even if the other planets of other systems are producing AMLs but a minute or two ahead . . .

This may not be as depressing as some people may see it. Regard the planetary systems produced by a supernova as a kind of family. This means that our parental exploding star could have ensured that our local area of the Galaxy has been and may still be under intense scrutiny from expanding spheres of vNs.

There is, of course, the possibility that we are the first in this part of the Galaxy, perhaps even the first AML in the entire history of the Universe. These are again *ad hoc* hypotheses and disprovable only by extensive ETIs.

Evidently the science policy makers in the SF scenario above were rapidly coming to the conclusion that we were alone by deciding to leave decision-making on the subject to the computers. They may not have made this choice consciously, but this is what it amounts to.

The last words I put down for Jess, the Press Secretary, are the whole point of the little tale. No one so far has thought about what we do once we discover ETI. It seems folly to me that governments throughout the world are considering investing millions into CETI without making more than the most basic steps to consider what the consequences of an encounter are likely to be.[2] All that we ever hear are individual opinions. Many entreat us to leave alone the whole possibility of contact with other intelligences but the alternative voices seem more concerted in their efforts and the public seems increasingly interested in what they have to say. Having a preponderance of literate popular figures in the CETI camp does help, of course.

It is crucial that we review and continue to review what the prospects of CETI are likely to mean to the human race. The odd chapter in a book, a couple of ill-attended hours at a CETI conference, and a dusty report or two over a decade old are just not good enough.

There are, basically, two types of encounter, so it should be fairly straightforward examining these and our response to them: the types are full encounter and transient encounter. Of these two there are two simple variants: active and passive. Let me provide some examples:

1A puts us in permanent close contact with ETI. Establishing communications with inhabitants of UFOs would fall into this category but I believe that opening a discourse with the local vN 'college' will be the most likely achievement in this category.

1B would be the discovery either of remains of ETI visitation to the Solar System or of a databank left for us by them in some bygone era. Inactive spaceprobes would fall into this category.

2A should be something like our setting up a temporary discourse with an intelligent starprobe which was passing through our own system without decelerating. They could actually be doing this frequently right now and we would be unaware of it.

2B might be a short burst of ETI radio signals which we picked up fleetingly before they disappeared, but were unable to decipher.

Radio CETI—that is, actual communication exchanges—would fall into category 1A whereas SETI, search only, would place any signals in 1B should we decide not to respond.

What about our own spacecraft encountering beings when they arrive at distant systems? Again using the cosmic-clock analogy we are liable to be a second or so ahead of our present stage when this occurs. One hopes that we will be deeply involved with the variety of vN we have discovered and will not try playing God with AMLs which have not achieved capabilities comparable with our own. May interstellar distances please act as quarantine to keep them safe from us until our minds are suitably overhauled.

The problems which might arise on our encountering another AML which was virtually matched with us in development are hair-curling. Behavioural anomalies on either side are likely to be viewed with disfavour by the unenlightened eyes of the other party. Given a high level of technology and a low level of responsibility someone is likely to make a step which we would all regret. Luckily it seems virtually impossible that we are going to encounter anything near our own level apart from vNs if even they are there.

Nevertheless, the outside possibility that we become involved in an encounter, even with a vN, which is potentially dangerous is reason enough for us to start thinking about xenethological theory right now. Admittedly, it is an area of pure speculation: nobody can guess with the remotest pretensions at accuracy what behaviour is likely to result from extraterrestrial environments.

There is one attitude which we must dispel from the start. It is one which has been beloved of SF movie moguls and writers since the days of H. G. Wells and can be suitably illustrated in this quote from *War of the Worlds*:

> Yet across the gulf of space, minds that are to our minds as ours are to those of the beasts that perish, intellects vast and cool and unsympathetic, regarded this earth with envious eyes and slowly and surely drew their plans against us.

Certainly the most obvious and most frightening consequence of encounter is that the human race may be annihilated. This is rather a peculiar way of looking at the whole subject. It reminds me of the peculiar attitude of some patients in mental institutions that everyone in the world is plotting for their destruction or that the safest place to stay is in bed. Going out of doors might lead to all sorts of disasters from being struck (either by lightning or a motor vehicle), to being robbed, raped, murdered or catching a disease from some dirty person you might pass in the street.

There is a more refined version of this which is that we will suffer an insuperable ego blow should beings vastly superior to ourselves make themselves known to us. The great danger here is that we will have nothing left to discover, all the vistas will have already been explored and there will be nothing left for Man to do but vegetate miserably. Apparently the same fate would then befall all of us as that which overtook the Eskimo when North American big business and Westernization in general destroyed the traditional way of life.

The idea of ego blow is really the largest segment of unmitigated excrement I have come across in my years of reading in the field of CETI. There are many points at which to begin and rave on about just how fatuous it is, but let me start thirty thousand years ago. The way of life of the hunter-gatherer peoples existing at the end of the last ice age has all but gone. It is in the nature of life on Earth and in the nature of human nature that the more successful tends to replace the less successful where they are in competition. The Eskimo way of life may well pass, regrettably, within the next few decades; this does not mean that those descendants of the Eskimo peoples cannot adapt successfully to other ways of life, and there is a considerable diversity of lifestyles available to individuals in the developed countries of the world today.

The second point, and the really salient one, is that our world today is a thriving mixture of peoples with varying types and different levels of economic, social and technological enlightenment. The historical examples of Japan and China in their response to a cultural and technological challenge over the last century show ego blow as the inane concept which it basically is.

Finally, those individuals who believe that there may come a point when any AML really can know everything there is to know have a very different concept of the Universe from that held by most philosophers and scientists today. If we do encounter a non-omniscient ETI then we will be in a position of having vast realms of knowledge and perhaps technological application to bring to bear upon them, the uniquely human viewpoint. When we meet a superior ETI we are unlikely simply to throw up our hands in despair, sit down and die. Certainly there are human beings who do this. If they were in the majority the species would be finished within the next few centuries whether or not we bumped into the little green men.

Historical example, as I have shown earlier in these pages, indicates that isolation as policy does not tend to work. In Japan and China the forces which ended isolation were as much internal as they were foreign. Even if the reverse zoo hypothesis proves to be the case at some point in the immense spread of the future, some humans, whatever their mental and physical form, are going to attempt to join the Club.

The 'leave them alone' school, the 'Zealots' as Toynbee would call them, will also raise the hue and the ringing cry when we discover vNs in the Solar System, as I feel sure we may do well within the next century. They will point out that we should not investigate such objects (leave them alone and they might go away or perhaps simply ignore us). Even if we just want to poke around with an old inactive probe they will come up with objections. In this case it is fairly obvious what the whine is likely to be: the technology which will be acquired from the probe is the product of a more mature culture and we are not yet ready for it.

If the probe is the product of a technological lifeform whose capabilities are some century to a century and a half ahead of us, then (if we discover the probe around the year 2000) it will bear the same relation to us as would an abandoned broken-down radio telescope to an astronomer of the mid-nineteenth century.

There are three problems which face anyone confronted with a mysterious piece of equipment. They will not be at all easy to overcome. The questions any aspect of a probe would raise are:

what does it do?; how does it do it?; how is it made?. The first two are the comparatively easy questions. They are causally related. The final question is the one which really will push forward the minds of scientists and engineers confronted with it. The analogy might be in the plans for an aircraft like the supersonic airliner *Concorde*. Anyone reading the technical press, and even some of the popular press, during the years of its design and development, would have come across reams of research papers, design diagrams and amazingly detailed plans. When a little piece of industrial espionage was indulged in by a nation which wanted to build its own SST what was stolen was what was unavailable—the details of how you *make* the functional product.[3]

Basically the point which I am making remains the same. It is going to be a day of some excitement whenever the encounter takes place. Even if it is just an inert junkpile drifting around the Sun the stimulus which it will deliver will be phenomenal.

The impact of the space race on the world's economy appeared from my uneducated viewpoint to be very positive, very strong and disseminating advantages to almost every country on this little planet. There seemed to be a big dumb booster sitting under the financial mausoleum of the 'fifties. Sputnik activated it. The economy of a country, never mind the world, is like the weather: there are a lot of experts but the damn' thing does what it wants to anyway.

If a handful of plastic and thin metal did that in 1957 the lift value of ETI encounter might raise the standard of living for the majority of the human race slightly above the poverty line. Hopefully that would just be for starters. This kind of potential financial thrust is surely reason enough for governments and industry to investigate the possibility of CETI as a worthwhile area for substantial investment.

It is interesting to note that there is an increasing tendency amongst businesses to look into the future. There are numerous industrial enterprises which invest heavily in market assessment and trend prediction. They frequently rely on the expertise of professional economists to build models of the future and see how invested capital is liable to grow under various conditions of political, technological and even meteorological change. Competing industries are matched as are competing corporations within the industries concerned.

All this is very healthy. NASA has recently brought before its critics a report indicating just what kind of contribution it is capable of making to the US economy alone.[4] The figures, although heavily criticized by the anti-space lobby, are substantial. The message should have been clear for some time now: space programmes pay off.

A CETI programme should also pay off.

Studies in this area should be initiated as soon as it is feasible to do so. Like now. The president in the little fictional introduction to this final chapter had no background knowledge with which to handle the encounter situation. This ignorance is quite unnecessary.

Apart from the possibilities of projecting how a CETI programme might influence the economy, let us also project what a successful encounter could do to it. Let us see what that encounter could do to us in other areas by running computer games similar to the complex sociological studies in which large-scale multifactor situations are modelled and allowed to change according to the patterns in which these models appear to change in real life. This way we will gain some valuable advance indicators of how the world will react to encounter.

There is, of course, the presentation of encounter to be considered. How do you tell the public and who tells whom what in the first place? Gaming encounter situations can only be helpful. The personal and personality dynamics concerned in a situation like this are important. How is the US public going to take the news from a president at an all time low in his popularity? In this case it may perhaps be advantageous to have a publicly highly rated well known newscaster to break the news on all the networks simultaneously. The number of strategies for tackling this event are quite diverse, and the more often they are gamed the greater will grow the body of proxy experience which should give some pointers as to which are likely to work fairly effectively, and the ones which will possibly be disastrous.

Furthermore, as we are moving now into an era when CETI is respectable, it is time that we started coming to terms with how we go about altering our ways of looking at things. Perhaps this is how we may make first steps in coming to terms with the ETI, when it becomes a reality. Again, we would be involved with a gaming principle. Some fairly sophisticated programs, with a lot of lateral thinking in their operation, would be required. In effect, what would be assembled in the computers would be mock AMLs distinct from our own. A telepathic/telekinetic ETI might be modelled, one with a very poor ability to think mathematically. How do we interact with this?

The trick is not to delude ourselves into believing that we are devising approaches that will provide an easy route to understanding AMLs other than ourselves. The value of this exercise is that we will be seriously approaching the problem as one which has non-human aspects. We will realize that these aspects are likely to be fundamental. The plus will come from having exercised ourselves. We will know more intimately than ever before what the limitations are to Man. Perhaps then we will be willing to change them.

Will such gaming have an advantage when run on an international basis? There is the chance that this is the one area in which the Warsaw Pact and NATO could contemplate a joint defence operation—planetary defence. This may help international relations by warming them from the chilly standoff climes which can precede the corpse-cold storms of active warfare.

This may not be a particularly bad thing. More emphasis would be placed upon cooperation in space and less upon confrontation on the ground. It would, of course, be inadvisable to let the military take over CETI. They may hit the first pinpointed vN with a missile or a high-energy pulse and vaporize it, out of sheer hysterical fervour.

Put the onus of the search upon international teams of civilians and let there be a permanent UN committee to oversee the project, make decisions and allot funds.

So much for the ideal. It may come about but what I envisage as more likely is a long heartbreaking radio CETI search which winds up eventually having no convincing results to its credit. When contact is finally made with vNs we will botch it. The cumulative effects of TV and comic book space invasions will bear fruit.

In the world of SF it would be so very heartening to find some philanthropic soul putting up an annual prize for the most responsible encounter situation in a comic book, TV show, movie or short story in any language. Some kind of move to rid ourselves of Wells' and Welles' Martians, once and for all, has to be made as a matter of public education. Likewise it would be encouraging to find a more enlightened attitude amongst the English-language press. The aura of crankishness must be dispelled.

An aid to dispelling this could be the founding of an institution which would meticulously inspect all supposed 'evidence' for UFOs and ancient ETI encounters. Such a body would hopefully not merely be a bunch of armchair bound disbelievers but would actually initiate search programmes for archaeological indicators and devise schemes for testing various UFO theories. The laudable North Americans who have formed the impressive Committee for the Scientific Investigation of Claims of the Paranormal might set an example here. Perhaps with some more generous funding they may even be persuaded to set up a similar open-minded body of impressive sceptics willing to look for ways of disproving their own hypotheses.

We must educate and be educated. We must dispel the arrogance and despair which are the products of believing both that we are alone in the Universe and that the Universe is planet Earth alone. Out there are living minds the likes of which are as beyond our limited imaginings as we are beyond theirs.

If we realize that we, like they, are a unique voice whispering in the dark a story which will never be heard again, then perhaps we will begin to see ourselves as precious, as all lifeforms surely are. Then perhaps we will begin to treat one another with the respect and compassion we so surely deserve.

Then perhaps we will begin . . .

Appendix 1

A Checklist of ETs in the Cinema

This is a selective compilation of films in which ET life is a primary ingredient. Those asterisked (*) are the ones which in my opinion have something to say about encounter situations. Comment is mostly excluded but there are some notes listed at the end.

Message From Mars, 1913.
A Trip to Mars, 1917.
First Men in the Moon, 1919 & 1964.*
Aelita, 1924.
Flash Gordon, 1936.
Flash Gordon's Trip to Mars, 1938.
Flash Gordon Conquers the Universe, 1940.
The Purple Monster Strikes, 1945.
Superman, 1948.
Rocketship XM, 1950.
Flying Disc Man from Mars, 1950.
The Thing (From Another World), 1951.
The Day the Earth Stood Still, 1951.*
Cat-Women of the Moon, 1951.
Man From Planet X, 1951.
Flight to Mars, 1951.
Radar Men from the Moon, 1951.
Red Planet Mars, 1952.*

It Came from Outer Space, 1953.*
War of the Worlds, 1953.
Invaders from Mars, 1953.
The Lost Planet, 1953.
Target Earth, 1954.
Devil Girl from Mars, 1954.
Forbidden Planet, 1955.*a
Beast with a Million Eyes, 1955.
This Island Earth, 1955.
The Quatermass Experiment, 1955.*c
Invasion of the Body Snatchers, 1956.b
Fire Maidens from Outer Space, 1956.
Twenty Million Miles to Earth, 1957.
Supersonic Saucer, 1957.
Atomic Rulers of the World, 1957.
Earth Versus the Flying Saucers, 1957.
Not of this Earth, 1957.
Kronos, 1957.
27th Day, 1957.*

The Giant Claw, 1957.
Invasion of the Saucer Men, 1957.
Quatermass II, 1957.
The Brain from Planet Arous, 1957.
Attack of the Fifty Foot Woman, 1958.
Astounding She-Monster, 1958.
War of the Satellites, 1958.
The Space Children, 1958.
The Blob, 1958.c
I Married a Monster from Outer Space, 1958.b
The Brain Eaters, 1958.
Night of the Blood Beast, 1958.
The Flame Barrier, 1958.
Space Master X-7, 1958.
Strange World of Planet X, 1958.
The Trollenberg Terror, 1958.
It! The Terror from Beyond Space, 1958.
The Mysterians, 1958.
The Cosmic Man, 1959.*
Invisible Invaders, 1959.
Plan 9 from Outer Space, 1959.
Teenagers from Outer Space, 1959.
Missile to the Moon, 1959.
The Angry Red Planet, 1959.
Battle in Outer Space, 1959.
12 to the Moon, 1960.
The Phantom Planet, 1960.
Journey to the 7th Planet, 1960.
Visit to a Small Planet, 1960.
Village of the Damned, 1960.*
Man in Outer Space, 1961.
Children of the Damned, 1963.*
Battle of the Worlds, 1963.
The Day Mars Invaded Earth, 1963.
The Day of the Triffids, 1963.

Unearthly Stranger, 1963.
Invasion of the Animal People, 1963.
Invasion of the Star Creatures, 1964.
Robinson Crusoe on Mars, 1964.
The Human Duplicators, 1965.
Frankenstein Meets the Space Monster, 1965.
The Earth Dies Screaming, 1965.
The Night Caller, 1965.
Mutiny in Outer Space, 1965.
Dr Who and the Daleks, 1965.
Queen of Blood, 1966.
Daleks—Invasion Earth 2150, 1965.
Invasion, 1966.
Arnas Infernales, 1966.
Destination Inner Space, 1966.
Mars Needs Women, 1966.
The Bubble, 1966.
Quatermass and the Pit, 1967.*a
They Came from Beyond Space, 1967.
The Terrornauts, 1967.
The Night of the Big Heat, 1967.
Destroy All Planets, 1968.
2001: A Space Odyssey, 1968.*a
Barbarella, 1968.
The Body Stealers, 1969.
The Illustrated Man, 1969.
The Astro-Zombies, 1969.
The Green Slime, 1969.
Night Slaves, 1970.
Slaughterhouse Five, 1970.
The Andromeda Strain, 1971.c
Yog-Monster from Space, 1971.
Solaris, 1972.*
Beware the Blob, 1972.
The People, 1972.
Phase IV, 1973.b

Dark Star, 1975.
UFO Incident, 1975.
The Man Who Fell to Earth, 1976.
Star Wars, 1977.

Close Encounters of the Third Kind, 1977.*
Starship Invasions, 1978.
Laserblast, 1978.

a Few SF films are imaginative enough to consider the fact that civilizations in the Galaxy may exist so far apart in time that they seldom connect on a 'live' basis. These ones do. They are all excellent encounter scenarios.

b These films come close to being asterisked. In each case I feel that the situation could have been more deeply explored, although admittedly this could have detracted from the storylines and the cinematic impact.

c The idea of a simple non-intelligent organism from space causing havoc is one which the cinema has left almost entirely unexplored. Two of the three films which have tackled it, *The Quatermass Experiment* and *The Andromeda Strain.* are to be commended.

Appendix 2

Relativistic Black Holes

One of the great tricks of SF is 'hyperdrive' or 'spacewarp' or whatever. The purpose of this mechanism is to bypass the restrictions of Einsteinian Relativity theory; sadly, this puts a rather heartless limit on the speed at which science-fiction writers can propel our imaginary starships through the Universe. 300,000 kilometres per second, the speed of light in a vacuum, is the ultimate in velocities, apparently.

'Hyperdrive' negotiates a way around this by placing the starship or some elements of it outwith Einsteinian spacetime. The possibility that something akin to it may exist has been raised by Professor John Wheeler of Princeton. His 'Superspace' does not contradict Einstein at all and is in fact firmly based on standard Relativity theory.

How do we find a way of utilizing Superspace?

This was one of the questions I pondered on when considering the possibility that ETIs may have discovered the appropriate mechanism, as we too may in the coming centuries. Could black holes be the answer? It has been postulated that they may be shortcuts through our own Universe or alleyways into another.

There are two ways of placing your spaceship in a black hole environment. The most straightforward is to bring them both together, the spaceship and the black hole. The other way is to turn your spaceship into a black hole.

The latter sounds just as difficult as merely finding Superspace in the first place, but this is not quite the case: in fact, it is a fairly simple process, in terms of Relativity theory at any rate. The clue lies in the fact that as a body travels faster it appears to become more massive. It also becomes shorter. The combination of these two factors means that, as a spaceship accelerates, it appears to become an increasingly dense object.

Black holes are objects which are so dense that, in crude terms, their gravity has crushed them out of contact with our Universe.

So could our ship ever become a black hole in the eyes of the Einsteinian Universe? It would seem so. To travel at the speed of light the ship would have to acquire infinite mass and therefore infinite density. This is out. Long before the ship reached infinite density it would achieve the density required for it to become a Relativistic Black Hole.

The term relativistic is the key. The ship would not become a black hole to the crew. It would merely appear to become one to observers on Earth.

Would it then have passed out of our Universe or found a short cut through it, or what?

The equation for the relationship between the diameter of a body and the mass it must have to take on the aspects of a black hole are known theoretically. This is:

$$D = \frac{4MG}{c^2}, \tag{1}$$

where D is the diameter, M is the mass, G is the gravitational constant and c is the speed of light. So what we are interested in is the point at which the relativistic mass of the spaceship is equal to the mass factor in the above equation. As we already have a figure for the diameter (D) of the spacecraft we can change the above as follows:

$$M = \frac{Dc^2}{4G}. \tag{2}$$

To relate this to the relativistic mass of the spaceship we should introduce the association between the ship's mass when it is stationary and its increased mass when travelling at relativistic speeds. As we are looking for the relativistic mass which gives black-hole density in relation to the diameter we will call the relativistic mass M, its stationary mass m and its velocity v.

The equation is found in the Special Theory and is

$$M = \frac{m}{\left(1 - \frac{v^2}{c^2}\right)^{\frac{1}{2}}}. \tag{3}$$

So, combining (2) with (3) we wind up with

$$\frac{Dc^2}{4G} = \frac{m}{\left(1 - \dfrac{v^2}{c^2}\right)^{\frac{1}{2}}} \; ;$$

$$\therefore \qquad Dc^2\left(1 - \frac{v^2}{c^2}\right)^{\frac{1}{2}} = 4Gm;$$

$$\therefore \qquad 1 - \frac{v^2}{c^2} = \left(\frac{4Gm}{Dc^2}\right)^2;$$

$$\therefore \qquad \frac{v^2}{c^2} = 1 - \left(\frac{4Gm}{Dc^2}\right)^2;$$

$$\therefore \qquad \frac{v}{c} = \left[1 - \left(\frac{4Gm}{Dc^2}\right)^2\right]^{\frac{1}{2}};$$

$$\therefore \qquad v = c\left[1 - \left(\frac{4Gm}{Dc^2}\right)^2\right]^{\frac{1}{2}}.$$

We have an expression giving us the velocity at which a spaceship, or anything else, becomes a black hole. Or do we?

The problem with all those pretty equations is that they are based on non-Einsteinian maths. The fact that equation (1) falls into the category of high school maths happens to be a sheer coincidence as far as anyone can tell. This said, let us try testing the observable Universe and see if we can come up with any answers or indicators.

First of all, we can assemble a picture of the relationships between known objects' sizes and masses and the black hole threshold.

In fig. 1 we can see that there is something very weird with the picture of the Universe that we might expect from our knowledge of black holes. Everything on the right-hand side of the Schwarzschild radius should be a black hole and thus invisible as light should, theoretically, be unable to escape it.

Quasars, however, appear to be sufficiently compact in size while somehow avoiding becoming black holes. Is there any way we can resolve this apparent paradox? What do we know about these distant objects of the Universe which relates to the RBH hypothesis?

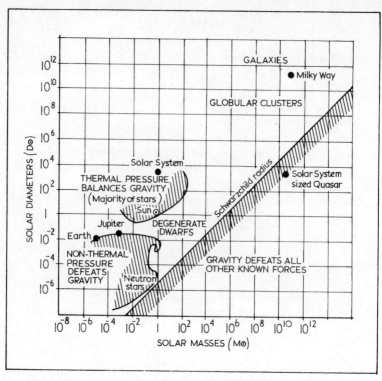

Fig. 1: The observable Universe in terms of solar masses and diameters.

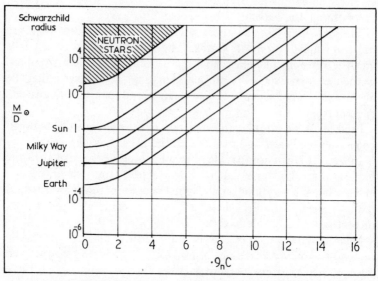

Fig. 2: Velocity and the Schwarzschild radius for a moving body.

We know that they are massive and that they are retreating from us at great velocities. Both of these make them candidates for scrutiny.

In fig. 2 the vertical axis on the left-hand side is a measure of objects' masses and diameters while the $.9_n c$ running horizontally represents the number of 9s after the point as a fraction of light speed (thus $.9_4 c$ represents $.9999c$). The shaded area for neutron stars in the top left includes stable neutron stars and those ones which are so massive that they are undergoing an irreversible collapse into the condition of becoming black holes.

Evidently, as soon as we start increasing mass relativistically among very distant galaxies we start to increase the number of black holes which they contain from our point of view. This should become evident at cosmological distances which imply recession speeds best expressed as fractions of light speed. The greater the fraction the more evident will become the presence of the relativistic gravitational effects. It should be readily observable in the bending it induces in the light coming from the distant galaxy concerned, as illustrated in fig. 3.

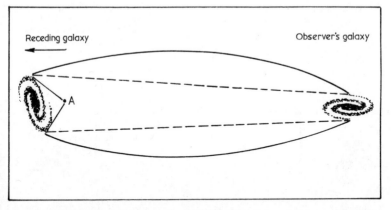

Fig. 3: Unbroken lines indicate actual path of light. This is curved between the rapidly separating galaxies but virtually straight for an observer at A who is so close to the observed galaxy that there is no cosmological recession. The broken line indicates the lightpath adopted in the absence of spacetime curvature induced by relativistic gravitational effects. Spacetime curvature induced by the Universe as a whole is ignored in this diagram.

Relativistic gravity induced by relativistic mass is bending the light which a distant galaxy is beaming at us. This means that the light will be reddened as it has to travel further—gravity has stretched it.

To determine the redshift, z, of a body travelling at relativistic speeds you must combine the relativistic doppler effect from Special Relativity with gravitational redshift from General Relativity. To estimate this you have to employ the difference between the rest mass M_o and the relativistic mass at the particular space-time event when you are measuring the body, M_e. This gives us the rather elaborate proposition

$$z = \sqrt{\frac{1+v/c}{1-v/c}} \cdot \sqrt{\frac{1-2GM_e/c^2r}{1-2GM_0/c^2r}} - 1. \qquad (4)$$

Now, if you turn back a page or two, you will see that from equation (1) we can talk about this in terms of a black hole where M is the mass of a black hole; we shall call this the Schwarzschild mass, M_s. It is convenient to regard r in (4) as a radius for the body which is half diameter D in equation (1).

What happens then to light reddening as the accelerating body increases in mass and thus gravity from an observer's point of view?

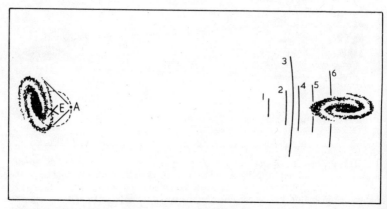

Fig. 4: E is a relativistic gravitational event observable at cosmological distances only by its effect on light wavecrests photons. Relativistic gravity generates such events with a source E at distances such as 1, 2, 3, 4, 5, 6 etc. A is too close for a relativistic gravity event at this distance (broken line) but if A were moving fast enough it would occur and the visual result would appear to happen to the image of the Galaxy as a whole very rapidly, creating a temporary distortion.

As M_e approaches M_s, redshift z should itself increase infinitely and ultimately become inaccessible to any system of measurement.

A further interesting phenomenon would be the influence of gravity 'events' induced by this effect; e.g., white dwarfs and pulsars becoming relativistic black holes. The wavefronts of these 'events' would be induced so far from their source that they would appear to influence the entire galaxy over a very short space of time.

As quasars appear to satisfy many of the above requirements I shall venture the postulate that they are in fact objects of galactic dimensions, perhaps fairly ordinary galaxies. They are so distant that they are visually distorted and endowed with apparent abundances of energy arising from the RBHs induced by their rapid cosmological recession from us.

There is an interesting rider to this in reference to Olbers' Paradox. If quasars are more or less ordinary galaxies then the Steady State theory or something approximating to it would have to hold true. In this case we would have the paradox which (more or less) asks why the night sky is dark when there are so many billions of galaxies with billions of stars belting out all that light. The two current explanations are: firstly, at great distances the light is redshifted out of the visual spectrum (in which case what about the ultraviolet which is then supposedly redshifted into it?); and, secondly, most objects in the Universe are so distant that their light has not yet reached us (in which case the Universe is not infinite in time). The alternative explanation which I put forward is that the stars, and eventually the galaxies themselves, become Relativistic Black Holes from which their light simply cannot escape to us.

Having happily worked out that my spaceships might escape the restrictions of lightspeed by turning themselves into RBHs I set about trying to figure out just how it would work out in fact or, rather, in science fiction. The results left me with an interesting hypothesis but some rather awkward angles to negotiate. Fig. 5 is an illustration of what I mean.

Accelerating at one g all neutron stars after some five hundred or so years have become Relativistic Black Holes, as have the majority of other stars by seven thousand years, Earth-type planets by twenty million years. So what about my starship, assuming it weighs about five thousand tonnes and is about twenty metres in diameter? It is powered by the wondrous Boycian Field Propulsion System (patent pending) at one g. Sadly, it would appear that it

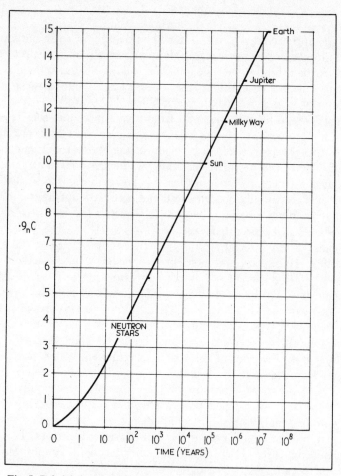

Fig. 5: Relativistic velocities attained by 1g acceleration against given time.

takes rather a long time to become an RBH by this method. It takes literally many hundred million billion billion billions of years. The trouble lies in the fact that the velocity, mass and time involved work out as expressions of the energy involved in an object becoming a black hole. To circumvent time I will accelerate at many hundred million billion billion billion g and pop out of our Universe in less than one year.

Needless to say, if I want that kind of energy on tap my spaceship will have to be powered not by nuclear reaction (even matter annihilation will not suffice) but by the only possible source—a black hole.

Well, this is where we came in . . .

Appendix 3

The Impact of vNs upon Material of the Solar System

Although it is impossible to imagine quite what particular matter would be needed by a vN in the processes of maintenance and reproduction we can still estimate the mass which should have been used up by them.

I will assume that they have been arriving ever since the Solar System was formed out of chaos some five or so billion years ago, and have been refurbishing and replacing themselves on a regular basis ever since, using materials gleaned from the System itself. This would be material either available in low-g areas like the smaller moons and perhaps the asteroids, or mined straight out of the Sun in the form of plasmas.

Where M is the mass of the vN, T is the lifetime of the System, R_1 the rate at which vNs appear in the System and R_2 the rate at which they reproduce we have an estimate for their mass in total from the formula

$$\frac{T^2 M}{2R_1 R_2}.$$

Knowing the approximate figure for asteroid density (roughly 2.5 tonnes per cubic metre) it is possible from this to work out the volume of an asteroid which would represent an equivalent mass. This is not to say that all requisite materials will be found in one area. It is merely a convenient way of regarding the impact which vNs will have made upon our System in general.

As I mentioned in Chapter 8 (see page 122), vNs will very probably scavenge their own kind in a process akin to cannibalism in order to acquire parts, so the figures given here are the upper limit that we can expect.

The table below gives two times numbered in years: T_1 is the average number of years between arriving vNs and T_2 represents the time lapsing between complete renewal of the vN complex the mass of which is given immediately below. The diameters listed under the T columns represent the diameters of asteroids equivalent in mass to the amount of matter used by those vNs over the past five billion years.

	T_1	200	500	1000	2000	4000	
	T_2	500	1000	1000	1000	1000	
Mass = 1000 tonnes							
Diameter equiv.		408	240	189	150	120	kilometres
Mass = 10,000 tonnes							
Diameter equiv.		880	515	408	325	258	kilometres
Mass = 100,000 tonnes							
Diameter equiv.		1896	1109	880	699	554	kilometres
Mass = 1,000,000 tonnes							
Diameter equiv.		4085	2389	1896	1506	1195	kilometres

These diameters vary from those of the satellites of larger planets to those of known asteroids. In an extreme case, say, the arrival and duplication of vNs each with the mass of a naval cruiser —say, twenty thousand tonnes—every decade, the diameter would be some eleven thousand kilometres, almost the size of our own world. This is still vanishingly small against the mass of the Solar System (about 0.0003% in fact).

Appendix 4

Tomorrow and Tomorrow: A Glance at the Future

When we are confronted with indisputable evidence for ETI we may have to face a society which was once like our own but has advanced considerably on our comparatively primitive condition. What is more likely is that our society will at some point in the future face a very unusual form of AML originating off the Earth. In either case, we should take a look at the future to imagine the possible contexts.

I will run through some projected developments for the next couple of centuries. I will be drawing heavily on three particular works for this: *The Next Two Hundred Years* by Herman Kahn; *Profiles of the Future* by Arthur C. Clarke; and *Tomorrow's World* by Raymond Baxter.

By 1985: Clandestine proliferation of nuclear weapons. Spectacular multinational degradation of the environment. Sophisticated surveillance and control of public behaviour. Widespread unemployment rising with inflation. Choice of sex for babies becomes moral issue. Cannabis and other mood-influencing drugs more readily available and less unacceptable. First child born outwith the womb. Large-scale internal warfare in Africa has begun. Regional technological catastrophes.

By 1990: Human cloning is possible. Highly efficient electrical storage systems. Cancer conquered. Permanent prosthetic devices grafted into human bodies. Brain enhancement by drugs. Fusion becomes viable. Cheap synthetic food.

By 1995: Cold cure and other anti-viral measures. Tremendous increase in wealth over past two decades particularly in Third World but proportional differences in wealth distribution. Nuclear terrorism widespread. EEC reshaping into a new Euro-community dominated by France. Possible estrangement of Eastern European countries from USSR and formation of their own trading/security grouping. New arms races as some Third World countries such as Brazil become major suppliers of arms in competition with 'developed' countries. USSR becomes increasingly subjectively oriented as domestic problems multiply. Commercial fusion operational as an energy source. Global computer links for home computer terminals.

By 2000: Inflation ('the poverty of affluence') mitigates against the cutting of the already record levels of unemployment. Minority agitation groups outlawed. High surveillance of ethnic and cultural minority groups as well as banning of many fringe political activities in the 'developed' countries. Two-way Brain-Drain where cheap-labour unemployed come to work for low wages in 'developed' countries and are replaced in the Third World by high-skilled unemployed from the West. Gross National Product averages out at $1000 per capita worldwide, but the figure disguises vast inequality of distribution. Tactical nuclear weapons have been used in limited warfare. Erosion of many traditional values among the educated throughout the world. Voluntary euthanasia. Sub-ice city in Antarctica. Genuine artificial intelligence. Full interface established between the human brain and the computer. Manned expedition to Mars. Permanent lunar stations and large orbiting space stations with workshop facilities. Beginnings of space industrialization. Alliance of 'developed countries' in Fortress North against developing Third World. Some countries adopt compulsory birth control. More food, including a good proportion of synthetics, becoming readily available throughout the world. Sea and seabed mined for minerals.

By 2025: Interstellar probe under way. Orbital power stations. Large-scale space industrialization. Santa-Claus Machines, self-reproducing factories which can manufacture almost anything when given the specifications. Complex issues being handled increasingly by humans with machine-enhanced brains. Computer-linked sensory extensions of the nervous system. Commercial fusion widespread as are non-depleting energy machines. Conservative backlashes developed to the erosion of traditional values. Emergence of 'educated incapacity' among post-industrial intellectual élites producing illusioned, irrelevant, ideologically exotic and impractical argument and attitude. Mutually conflicting cultural developments appear with a world-wide spread, resulting in 'mosaic cultures'. Crumbling of Fortress North as a viable concept. Law and order become a global problem. Possible emergence of agencies like Orwell's Thought Police. Genetic choice of child's general characteristics. Experimentation in nuclear synthesis of materials. Multicorp combines and alliances wield greater power than any national government.

By 2050: Virtually all energy comes from fusion and non-depletable resources. Design and construction of vN. International commuter system on land/sea/underground operating supersonically. Fully integrated man-machine systems tailored to one another appear. Average IQ rises dramatically world-wide. Human-artificial intelligence Hybrids take on many major decision-making tasks. Explosive colonization of space taking place. Gaps between wealthy and not-so-wealthy narrowing noticeably. 'Educated Incapacity' becomes major stumbling block to problem-solving due to overemphasis on theory much of which has no practical application, and even where it has, there is not enough work experience to fall back on. Widespread use of electrochemical mind enhancement, memory transplanting and relaying, and electronic telepathy as a communication device. Nuclear synthesis of materials operating commercially. Animals with genetically raised levels of intelligence. Experiments with gravity control.

By 2100: Political powers are (1) Multicorps based in Inner Planet space, and (2) Multicorps based on Callisto and dominating asteroid belt. Human race clearly splitting into two groups, the humans and the Hybrids. Latter have launched first interstellar expedition. Terrestrial food supplies greatly increased due to factory-produced synthetic foods, reclaiming of land from inhospitable areas, desalination of sea water and sea farming. Total weather control is in sight. Santa-Claus machines provide wide ranges of 'free' goods. Individual humans becoming more involved with communities and small-scale events and dissociating from the organization society. Mechanical instant education. Global computer teaching-library. Hybrids form first real think-tank by joining to formulate first true superintellects. Astronomical engineering projects are possible. Hybrid immortality is assured.

By 2150: Humans are on the wane as population begins to drop dramatically. Hybrids undertake large-scale social engineering of humans. A new hybrid—the Hybrid/dolphin hybrid—achieves the long-awaited comprehension of the cetacean mind. From this comes enhancing of the cetacean brain and links thereto. Differences in wealth are still evident but no longer important. New taboos and intellectual cults abound. Politics of the industrial societies of the two previous centuries is virtually obsolete. Politics is more community-oriented. Traditional states exist as technicalities. Multicorps employ virtually no one and are run by Hybrids with vanishingly small human involvement.

By 2180: Social engineering, the excising of unwanted characteristics, is almost completed amongst both Hybrids and 'humans'. Immortality for 'humans' assured. Hybrids begin world laboratory experiments on Venus.

Most of the above is my own interpretation of the sights seen in the smoky mirror and should not be laid at the doors of others. The other alternative futures can be summed up fairly briefly:

(a) Thermonuclear war: a massive debacle which plunges us back well over a hundred years at worst but at best results in merely a temporary halt of a few years (or perhaps even an incentive to increase in peaceful nuclear technology).

(b) Environmental catastrophe: the regional catastrophes which will become common before the end of the century will not have set up a sufficient control backlash to avoid a terminal situation again resulting in massive death and destruction. At worst this might set us back a hundred years, if we survived at all. Energy crises becoming explosive situations could lead to either this or (a).

(c) Graduated growth: this entails heavy restrictions to modify social impact and extends the above scenario of two hundred years by a further century.

(d) Zero growth: small-scale technological society with very low growth rate. Altruistic anarchism is the main political base for social structures. Extend above scenario by five hundred years to infinity. No crises other than on the very small scale should arise.

Notes

Prologue

1. Alvin Toffler is the author of *Future Shock*, a book which all students of ET Encounter situations should read. Toffler is concerned with the rising rate of change to which the individual is becoming exposed and the inadequacy of most to cope with it.
2. *Glasgow Herald*, 21 November 1977, 'Inside America' column, p 2.
3. 'Second BIS Conference on Interstellar Travel and Communication', K. W. Gatland, *BIS Journal* vol 30, no 12, December 1977.
4. This, of course, is how the Sun does function. Hans Albrecht Bethe demonstrated this satisfactorily in 1938.
5. Two editions of Koestler's *Act of Creation* have appeared. This quote comes from the earlier non-definitive one.
6. *The Psychology of Imagination*, Jean-Paul Sartre, p 20, Methuen, London, 1972.
7. *Ibid*, p 60.
8. *The Hollow Men*, T. S. Eliot, 1925.
9. *The Cosmic Connection*, Carl Sagan, p 42.
10. Extraterrestrial vampirism goes back to Wells' *War of the Worlds*. Here the Martians ingested nourishment in the form of human blood directly into their own bloodstream. The movie in which a green lady puts the bite on us Earthies is *Queen of Blood*.
11. This film is a prime example of one particular Hollywood SF cliché. Human beings are replaced with duplicates similar in all respects but for the absence of any emotions; a reflection of Wells' 'cool, unsympathetic intellects'.

Chapter One   Historical Perspectives 1

1: *Homo erectus* of a million or more years ago would certainly have had the basis for cultural exchanges. These people were stone technicians and as such had the basis for the interchange of technology. Designs of chisels, axes, anvils, etc., as well as novel techniques in using them would probably be valuable in such communities. There may even be evidence to suggest that cultural activity could have pre-dated *Homo* himself. See 'Evidence for Social Custom in Wild Chimpanzees' in *Man* vol 13, no 2, pp 234-251.

2: This is a very interesting if highly personal interpretation of history. It is valuable for the highlights which it offers on matters arising out of intercultural contact. Toynbee is generally regarded as an 'historicist', one who seeks to show general laws operating within the process of history. Karl Popper, the philosopher, argues strongly against this interpretation of history.

3: Strangely enough, no one has used the analogy with respect to what appears to be the worst example of destructive intercultural consequences. The natives of Tasmania suffered a population collapse when exposed to the West, to Britain in fact. Within one generation the 30,000 native inhabitants of the island were reduced to a few hundreds. They are now extinct.

4: The notion that European technology was totally alien to the native Americans is false. In the early struggles along the Americas' North Atlantic seaboard the natives proved this decisively. Not only did they learn how to run ships-of-the-line, on one notable occasion they displayed their command of naval tactics by soundly defeating British warships in a maritime engagement (1724, off the coast of New Hampshire).

5: It is only more likely because we can now be heard in terms of radio in an area comprising less than one hundred and twenty-five billionths of the Galaxy's volume whereas, before, we were heard over still less.

6: This transformation was mainly carried out within the fifty years from 1855 to 1905, although some fundamental rethinking did take place between 1945 and 1948/50.

Chapter Two   Historical Perspectives 2

1: *The Neophiliacs* by Christopher Booker is a study of the personalities and attitudes which made the swinging 'sixties swing.

2: No doubt gimlet-eyed Scots Nationalists will be scrutinizing any possible bias, for example the setting up of a St. George's Society.

3: *The Ascent of Man*, Bronowski, p 437.

Chapter Three   The Dark Pedigree

1: The Encounter Equation is not strictly mathematical but, if we are to regard it as an equation at all, it is best to establish some form of mathematical expression of it. We will represent the complex function Man as M and the complex function which is the ETI as E; human aspects not shared by the ETI we shall designate x and ETI aspects which are not shared by humans we shall term y. This gives us an 'encounter equation' as follows:

$$M - x = E - y.$$

$\therefore$ where $x = 0$ *or* $y = 0$, then $M < E$ or $M > E$, respectively, in terms of mental capability. Where both x and $y = 0$ then $M = E$; we are virtually identical with the ETI in terms of mentation.

2: A list of the forty-two interstellar molecules discovered up to early 1977 can be found on p 269 of *The Cambridge Encyclopaedia of Astronomy*.

3: Life is estimated to have existed in some form or another on Earth for more than three billion years; however, the pre-biological processes which produced terrestrial life may also have been singular. If this is so, then we can expect the fundamental structures of organisms to have differing 'architecture' even if they have the same 'building blocks'.

4: Here I am referring to the excellent examination of the rise of human technical civilizations by Richard Lee of Toronto University's Anthropology Department in *Communication with Extraterrestrial Intelligence: CETI*, edited by Sagan. This tells us a great deal about ourselves but from ourselves alone we can say nothing about other mental life in the Universe. We can certainly not assume that other mental life has civilization or technology. All we can say is that if they have technology as we appreciate it there is a chance that they have civilization, and vice versa. What the chance is nobody knows. Generally it must be regarded as suspect to assume that in the Encounter Equation (see Note 1 above) it is likely that x will be near zero.

5: These experiments in simian language capabilities took place respectively at the University of Nevada and the Yerkes Primate Center, Emory University, Atlanta.

6: There is a rumour afoot at the time of writing that an optimistic conservation group intends setting up just such an embassy in Sydney harbour.

7: Ian Watson's *The Martian Inca* is particularly interesting in this context; and his *The Jonah Kit* has an especial interest in interaction of human and cetacean thought processes.

Chapter Four   Come Up and See Me Sometime
1: The term 'biogrammar' was coined by Fox and Tiger and used in their joint work *The Imperial Animal*.
2: The hypothesis that all organisms interact in such a way as to maintain Earth as a fit habitat for life is called the *Gaia* hypothesis after the Greek goddess of the Earth who gave birth to the sky, mountains and sea. Its concept originated with James Lovelock, an analytical chemist, and has since found another champion in the person of Lynn Margulis, a cell biologist at Boston University.
3: *Star Trek* was selected here in preference to the great number of other TV science fiction series because it was consistently better than most others and, where others *were* better, they did not have as large an audience.
4: Of course there is no way of knowing the motive behind an ETI's behaviour on first contact; consequently it will be impossible to judge just how effective whatever method of behaviour manipulation it chose might prove to be.
5: Of course there is always the remote (I hope) possibility that instant panic will prompt us to send a hailstorm of nuclear warheads out upon the visitor.

Chapter Five   A Meeting With Tomorrow
1: Teilhard de Chardin, *The Phenomenon of Man*.
2: Here I must mention the Committee for the Scientific Investigation of Claims of the Paranormal. This is an organization of highly qualified sceptics, exactly the kind of body which the pro-paranormal community needs. They provide novel and always relevant methods whereby paranormal claims are disproved. Only when they have been satisfied that the paranormal exists will the rest of society really take it seriously.
3: This is not intended in any way as a teleological statement. Our 'direction' may change causing us to head towards a different 'point'. There is nothing to suggest that any point on the horizon is the destined end towards which we are being drawn or driven. In evolutionary terms it is better to regard the horizon as we regard the horizon in the world about us: it is merely a limit on our vision, and points exist beyond it which we cannot see.
4: Like H. G. Wells' *First Men in the Moon* this book examines extraterrestrials which have a biological caste system rather similar to that of the social insects. It is an extremely interesting investigation of this type of society.

5: The imaginary wars portrayed in this and other works are critic-
ally considered in I. F. Clarke's *Voices Prophesying War 1763–1984*.
6: Although this book is concerned with the future of Man rather
than the ETI it is probably one of the most effective US novels
produced since the Second World War. I cannot recommend it
highly enough.
7: Wiener, a distinguished mathematician, brought the word into
common currency when he published his book *Cybernetics* in 1948.
The term is derived from the Greek word *kybernetes* for 'helmsman'.
8: For a survey of the arguments and history associated with this
tender topic see *The Human Pedigree*, by Anthony Smith.
9: *Datamation*, no 2, April 1972, p 8.
10: Whether or not such machines ever appear seems to depend
(at least presently) upon our deciding to permit their existence, as
it is we who must set the self-improving system in motion primarily.
11: Man appears to employ tools to help him accomplish pre-
determined ends. For this reason I see the amplification of human
intelligence by direct interface with machines as inevitable, but
artificial intelligence as merely possible.
12: J. D. Bernal and Columbia University physicist Gerald Feinberg
have, however, speculated to some extent upon it and concluded
that, as the network of minds can be amplified, no mind need ever
be totally lost, even through death!

Chapter Six   Saucers and Dishes

1: Anyone quite new to the subject is particularly directed to Ian
Ridpath's two excellent introductions—see the Bibliography.
2: *The Cosmic Connection*, Sagan, p 203.
3: It really is important to state that at the time of writing there is
no reason to believe that there were advanced technical human
civilizations in the past (i.e., capable of handling energy in the
quantities which we can today) or that we have ever been visited
by ETIs. All evidence in support of such claims has been at best
ambiguous.
4: This was discovered by mining engineers in the Gabon a few
years back. Luckily for them it had been extinct for several million
years. See *Encyclopaedia Britannica* supplement 1978.
5: M. W. Saunders mentions the possibility of living databanks in
his 'Databank for an Inhabited Extra-Solar Planet'. This is a survey
of databank options open to a human interstellar expedition. *BIS
Journal*, vol 30, pp 349–358.

6: Shklovskii's speculation that the Martian moon Phobos might be artificial was mistaken but a fascinating description of the possibility can be found in *Intelligent Life in the Universe*, Shklovskii and Sagan, pp 363-376.

7: 'Communications from Superior Galactic Communities', *Nature*, vol 186, 1960, p 670.

8: The BIS Daedalus study is particularly interesting in this respect. It examines the possibilities of sending a probe some two parsecs from our Solar System to that of Barnard's Star using existing and foreseeable technologies.

9: Published in English with Carl Sagan's contributions as *Intelligent Life in the Universe*.

10: To be more precise, he is seeking out the characteristic radiation which would be emitted from a Dyson sphere or shell.

11: It has been suggested by P. V. Makovetski, of the Leningrad Institute of Aviation Instrument Manufacture, that perhaps the best plan for detecting a technical civilization would be investigating wavelengths which were significant in terms of mathematical and physical constants (such as π and the square root of 2). 21cm divided by π, for example, is 6.68cm.

Chapter Seven   Past Contact and the Moving Caravan

1. Duncan Lunan, *Man and the Stars*, Souvenir Press, London, 1974. US edition: *Interstellar Contact*, Henry Regnery Co., Chicago, 1975.

2. A. T. Lawton & S. J. Newton, 'Long Delayed Echoes: the Search for a Solution', *Spaceflight*, vol 16, no 5, pp 181-187 (May 1974).

3. C. G. Jung, *Flying Saucers: a Modern Myth of Things Seen in the Skies*, translated by R. F. C. Hull, Routledge & Kegan Paul, London, 1959.

4. Christopher Evans, *Cults of Unreason*, Harrap, London, 1973.

5. Carl Sagan, *The Cosmic Connection*, Doubleday, New York, 1973.

6. Zdenek Kopal, *Man and His Universe*, Rupert Hart-Davis, 1972.

7. Edward Purcell, 'Radioastronomy and Communication through Space', in *Interstellar Communication*, ed. A. G. W. Cameron, Benjamin, New York, 1963.

8. 'Eye on the Future', *Nature*, vol 241, 1973, p 363.

9. Eugene F. Mallowe, review of *Interstellar Communication: Scientific Perspectives*, *Journal of the British Interplanetary Society*, vol 28, no 3, pp 223-224 (March 1975).

10. Gerald M. Webb, review of *Interstellar Communication: Scientific Perspectives, Spaceflight*, vol 17, no 4, p 158 (April 1975).

11. Ronald Story, *The Space-Gods Revealed*, New English Library, 1977.

12. I. S. Shklovskii & Carl Sagan, *Intelligent Life in the Universe*, Holden-Day, 1966.

Chapter Eight Meditations of the Heart

1: 'Advanced life in the Universe', by A. Cottey in *New Scientist* 27 April 1978, p 236.

2: Paul Davies, of King's College, London, makes this suggestion in *The Runaway Universe*, p 101. It is interesting to note that the concept of biological spaceships has been tackled by science-fiction writers, notably in James White's 'Spacebird'. There is also Larry Niven's interstellar lifeform the starseed, which appears in 'Grendel'.

3: This, to be fair, is generally regarded as the upper limit for the possible emergence of Advanced Mental Lifeforms.

4: I first came across this concept in Michael Arbib's article 'The Likelihood of Evolution of Communicating Intelligences on Other Planets', contained in Ponnamperuma and Cameron's *Interstellar Communication: Scientific Perspectives*.

5: John von Neumann died in February 1957. The last year of his life was a tragic race against the bone cancer which agonized and exhausted him. It was a race to try and complete as much work as was physically and mentally possible by a man who had so much to give. An account of it by his wife can be found prefacing his last (unfinished) work *The Computer and the Brain*.

6: The BIS Daedalus study shows the promise of pulsed nuclear fusion for propelling interstellar vehicles. At the time of writing we are still several decades removed from the point at which we will sit down to the actual designing of such a craft; consequently we can imagine that numerous other and perhaps better propulsion system possibilities will become more readily accessible to Man.

7: Theodore Taylor of the International Research and Technology Corporation theorizes that a 'Santa-Claus Machine' might be possible, a machine which can literally build anything which it is programmed to build—including other 'Santa-Claus Machines'. Such a device would be very attractive for the core of a human vN system.

8: The most notable entries in this encyclopaedia can be found in Isaac Asimov's *Foundation* trilogy.

9: This can be found in Anthony Lawton's 'Long Delayed Echoes—The Trojan Ionosphere', in the December 1974 issue of the *BIS Journal*.

Chapter Nine Unbelievably Fatuous Observations

1: These are:

CE1K: no lasting trace of interaction between the UFO and the environment, nor with the reporting observer(s).

CE2K: observable traces of reported interaction remain subsequent to UFO report.

CE3K: reports referring to observations of animated occupants of UFOs.

See J. Allen Hynek's *The UFO Experience*.

2: Two interesting books which deal with gullibility and irrationality are Patrick Moore's *Can You Speak Venusian?* and Christopher Evans' *Cults of Unreason*.

3: An interesting examination of the evidence for fairies is in Robert Sheaffer's 'Do Fairies Exist?' in the Winter 1977 issue of *The Zetetic*. More recently, Summer 1978, the same magazine published an examination of fairy photographs using computer enhancement to illustrate their fakery.

4: See Ian Ridpath's *Messages from the Stars*, p 211 for a description of investigations by A. H. Lawson of California State University.

5: Jung's book *Flying Saucers: A Modern Myth of Things Seen in the Skies*, interprets the UFO phenomenon strictly within the terms of theoretical psychology.

6: This frightening thought process probably did occur. That official circles were more trepidacious of UFO reports than UFOs themselves is made abundantly clear in David Michael Jacobs' *The UFO Controversy in America*. P 83 specifically quotes a Robertson panel finding that such reports made the public vulnerable to psychological warfare.

7: This author's book *The Real World Of Spies* is a classic of espionage writing.

8: A report on this appeared in the *Inside France* column of the *Glasgow Herald*, 25 October 1977, p 2.

9: It may just be that Scotland is in the middle of a UFO 'spasm' coincident with my writing. Newspaper reports of sightings on or around my native soil rose sharply between summer 1977 and spring 1978.

10: See *Spaceflight*, June 1976, for Peter Glaser's 'Development of

the Satellite Power Station', and the same magazine for March 1977 giving the Boeing Company's concept considerations in 'Powersat'.
11: I am not alone in believing that we cannot yet dismiss the possibility of Martian life. Robert Jastrow, founder and director of NASA's Goddard Institute for Space Studies is also of this view. See his book *Until the Sun Dies*, pp 154–163.

Chapter Ten   Them
1: The stories of James White in this series alone should be required reading for exobiologists in training. He also has written a most interesting and instructive article on 'Biologies and Environments' in Brian Ash's *Visual Encyclopaedia of Science Fiction*.
2: Sneath's book *Planets and Life* probably gives the best introduction to the life-sciences aspect of the subject of ETI. It is a wide-ranging lucid work and comes highly recommended.
3: Royce's article 'Consciousness and the Cosmos' is contained in *Extra-Terrestrial Intelligence: The First Encounter* edited by J. L. Christian. This volume contains a number of valuable philosophical essays on the subject, notably Royce's and those of Peter Angeles and Wilfred Desan.
4: This is an aspect advanced by Wilfred Desan (see note 3).

Chapter Eleven   High Society
1: The story that Fermi, the great nuclear physicist, was supposed to have put this question over a luncheon in 1940 is possibly apocryphal, according to Carl Sagan in *The Cosmic Connection*, p 229.
2: *The Galactic Club* is actually the title of his excellent book on CETI.
3: This is probably the most exhilarating text in the entire field and not to be missed on any count by layman or professional!
4: This term was created by the Canadian communications philosopher Marshall McLuhan to describe the intimacy possible on a world-wide scale through supersophisticated electronic communications.
5: *The Ghost in the Machine* is Koestler's main treatment of this point.
6: Alan Cottey of the University of East Anglia warns of the danger of trying to separate the intelligence from its artifacts (Chapter Eight, note 1). This would certainly be the case for Man when he ever achieves this status. There will either be the situation where

we must view our artifacts as actual parts of the human or we must view what we have by then become as being human no longer.

Epilogue What Do You Say Before You Say Hello?

1 : How close the supernova would have to be is not quite certain but, if a supernova occurs within our Galaxy once every century (generally in areas obscured from us by thick dust clouds), there should have been one for every hundred or so stars. It takes several hundred thousand years for the shockwave to peter out and, after the first century, it is still expanding at some 10% of light speed. For further details see 'A Supernova Trigger for the Solar System?' by D. Paterson in *New Scientist*, vol 78, no 1102, pp 361–363, 11 May 1978.

2 : James Gunn's novel *The Listeners* should be compulsory reading for anyone involved in the funding of SETI programmes. It is a story about the first signal picked up from another star system after many decades of waiting, during which the project has degenerated into a minor financial /political chesspiece.

3 : Tom Norton, Conservative MP for Cheadle, made these claims with reference to Concorde machine tool manufacture and design in the European Parliament at Strasbourg. *Daily Telegraph*, 9 July 1975.

4 : The 1976 report commissioned by NASA from Chase Econometrics Associates Inc.; see 'Does Space Research Help US Economy?' by Roger Levis in *New Scientist*, vol 77, no 1096, p 838, 30 March 1978.

Bibliography

This is not a bibliography in the sense of a complete listing of works relevant to the subject matter of the book. My own feeling about CETI is that it is a subject of *fundamental* importance because so many diverse elements of human thought are involved here. What follows are two booklists. The first deals with non-fiction and the second with science fiction. The first is a selection of works which I hope illustrates the main arguments within CETI and the fields which I believe are essential to a study of it. The asterisked works might prove useful to newer readers in the field.

History:
Carr, E. H., *What is History?*, Macmillan, London, 1961*
Durant, Will, *The Story of Civilization* (32 vols), Simon & Schuster, New York, 1935, 1963*
Roberts, J. M., *The Hutchinson History of the World*, Hutchinson, London, 1976
Toynbee, Arnold, *A Study of History* (single volume edition), Oxford University Press with Thames and Hudson, London, 1972

Science and Philosophy:
Bronowski, J., *The Ascent of Man*, B.B.C., London, 1973*
Dobzhansky, T., *The Biology of Ultimate Concern*, New American Library, New York; Fontana/Collins, 1967
Duncan, Ronald and Miranda Weston-Smith (eds.), *The Encyclopaedia of Ignorance*, Pergamon Press, London, 1977
Koestler, Arthur, *The Act of Creation*, Hutchinson, London, 1964
Koestler, Arthur, *The Ghost in the Machine*, Hutchinson, London, 1967
Koestler, Arthur, *Janus, a Summing Up*, Hutchinson, London, 1978
Koestler, Arthur, *The Sleepwalkers*, Hutchinson, London, 1959*
Magee, Bryan, *Popper*, London, Fontana/Collins, 1973
Monod, Jacques, *Chance and Necessity*, Collins, London, 1972

Teilhard de Chardin, P., *The Phenomenon of Man*, Harper, New York; Collins, London, 1959

Teilhard de Chardin, P., *The Future of Man*, Harper, New York; Collins, London, 1964

Behaviour:

Ardrey, Robert, *The Hunting Hypothesis*, Collins, London, 1976*

Friedrich, Heinz (ed.), *Man and Animal*, MacGibbon and Kee, London, 1971

Leakey, Richard E. and Roger Lewin, *Origins*, Macdonald and Janes, London, 1977*

Lorenz, Konrad, *Behind the Mirror*, Methuen, London, 1977

Lorenz, Konrad, *On Aggression*, Methuen, London, 1966

Pfeiffer, John E., *The Emergence of Man*, Harper & Row, New York, 1969; Thomas Nelson, London, 1970*

Tiger, Lionel and Robin Fox, *The Imperial Animal*, Secker & Warburg, London, 1972

Wilson, Edward O., *Sociobiology: The New Synthesis*, MIT Press, Cambridge, Mass., 1975

Astronomy:

Asimov, Isaac, *The Universe*, Penguin Books, London, 1967*

Davies, Paul, *The Runaway Universe*, Dent, London, 1978*

Gribbin, John, *Our Changing Universe*, Macmillan, London, 1976

Jastrow, Robert, *Until the Sun Dies*, Souvenir Press, London, 1978*

CETI:

Asimov, Isaac, *Is Anyone There?*, Doubleday, New York, 1967*

Bracewell, Ronald, *The Galactic Club*, W. H. Freeman, New York, 1975*

British Interplanetary Society, Journal of, 'Final Report of the Project Daedalus Study Group', *Supplement*, 1978

Christian, J. L. (ed.), *Extraterrestrial Intelligence: the First Encounter*, Prometheus Books, Buffalo, 1976

Lunan, Duncan, *Man and the Stars*, Souvenir Press, London, 1974; as *Interstellar Contact*, Henry Regnery, Chicago, 1975*

Maruyama, M. and A. Harkins (eds.), *Cultures Beyond the Earth*, Vintage Books, New York, 1975

Ponnamperuma, Cyril and A. G. W. Cameron (eds.), *Interstellar Communication: Scientific Perspectives*, Houghton Mifflin, Boston, 1974

Ridpath, Ian, *Messages From the Stars*, Fontana/Collins, 1978*
Ridpath, Ian, *Signs of Life, the Search for Life in Space*, Penguin Books, London, 1977*
Sagan, Carl (ed.), *Communication with Extraterrestrial Intelligence: CETI*, MIT Press, Cambridge, Mass., 1973
Sagan, Carl, *The Cosmic Connection*, Anchor Press, New York, 1973; Hodder & Stoughton, London, 1974*
Shklovskii, I. S. and Carl Sagan, *Intelligent Life in the Universe*, New York, 1966; Pan Books, London, 1977
Sneath, P. H. A., *Planets and Life*, Thames and Hudson, 1970*
Stoneley, Jack and A. T. Lawton, *CETI*, Wyndham Publications, London, 1976
Stoneley, Jack and A. T. Lawton, *Is Anyone There?*, W. H. Allen, London, 1975
Sullivan, Walter, *We Are Not Alone*, McGraw-Hill, New York, 1964; Hodder & Stoughton, London, 1965

Future:
Bernal, J. D., *The World, The Flesh and The Devil*, (1929), Jonathan Cape, London, 1970
Berry, Adrian, *The Next Ten Thousand Years*, Jonathan Cape, London, 1974
Bundy, Robert (ed.), *Images of the Future: The Twenty-First Century and Beyond*, Prometheus Books, Buffalo, 1976
Calder, Nigel, *Spaceships of the Mind*, B.B.C., London, 1978*
Kahn, Herman and others, *The Next 200 Years*, Associated Business Programmes, London, 1977
Leach, Gerald, *The Biocrats*, Johnathan Cape, London, 1970
Rorvik, David, *As Man Becomes Machine*, Souvenir Press, London, 1973*
Toffler, Alvin, *Future Shock*, Bodley Head, London, 1970

UFOs and Ancient Astronauts
Hynek, J. Allen, *The UFO Experience*, Henry Regnery, Chicago; Abelard-Schuman, London, 1972*
Jacobs, David Michael, *The UFO Controversy in America*, Indiana University Press, Indiana, 1975
Klass, Paul J., *UFOs Explained*, Random House, New York, 1974*
Steiger, B., *Project Blue Book*, Ballantine, New York, 1976
Story, R. D., *The Space-Gods Revealed*, New English Library, London, 1977*

Temple, Robert K. G., *The Sirius Mystery*, Sidgwick & Jackson, London, 1976
von Däniken, Erich, *According to the Evidence*, Souvenir Press, London, 1978*

This second list is of science-fiction books. This is not intended as a guide to the literature. Anyone seeking such a guide would be advised to seek out such works as Brian Aldiss' *Billion Year Spree* (1973) for historical and philosophical viewpoints, and Brian Ash's *Visual Encyclopaedia of Science Fiction* (1977) for a good general introduction.

What follows is a selection of SF which should provide the reader with a wide variety of viewpoints on the ETI. It does not contain all the best ETI novels in the field and I have missed a fair number which other authors would certainly have included out of their own personal preferences. Regard it as a rough basic list. Most have gone into many editions—hence the lack of bibliographical details.

Aldiss, Brian: *The Dark Light-Years*
Aldiss, Brian: *Hothouse*
Asimov, Isaac: *The Gods Themselves*
Anderson, Paul: *The Byworlder*
Benford, Gregory: *In the Ocean of Night*
Blish, James: *A Case of Conscience*
Brunner, John: *Total Eclipse*
Clarke, Arthur C.: *Childhood's End*
Clarke, Arthur C.: *Rendezvous with Rama*
Clarke, Arthur C.: *2001: A Space Odyssey*
Clement, Hal: *Iceworld*
Clement, Hal: *Mission of Gravity*
Clement, Hal: *Needle*
Dickson, Gordon R.: *The Alien Way*
Green, Joseph: *Conscience Interplanetary*
Lem, Stanislaw: *Solaris*
Gunn, James E.: *The Listeners*
Herbert, Frank: the *Dune* trilogy (*Dune, Dune Messiah* and *Children of Dune*)
Hoyle, Fred: *A for Andromeda*
Hoyle, Fred: *The Black Cloud*
Lewis, C. S.: *Out of the Silent Planet*

Niven, Larry and Jerry Pournelle: *The Mote in God's Eye*
Shaw, Bob: *A Wreath of Stars*
Shaw, Bob: *The Palace of Eternity*
Stapledon, Olaf: *Last and First Men*
Watson, Ian: *The Martian Inca*
Weinbaum, Stanley G.: *A Martian Odyssey*
Wells, H. G.: *The First Men in the Moon*
Wells, H. G.: *The War of the Worlds*
White, James: *All Judgement Fled*
White, James: *Hospital Station*
White, James: *Star Surgeon*
Zerwick, Chloe and Harrison Brown: *The Cassiopeia Affair*

Index